Christianity for Doubters

Christianity for Doubters

GRANVILLE SEWELL

RESOURCE *Publications* • Eugene, Oregon

CHRISTIANITY FOR DOUBTERS

Resource Publications
An Imprint of Wipf and Stock Publishers
199 W. 8th Ave., Suite 3
Eugene, OR 97401

www.wipfandstock.com

PAPERBACK ISBN: 978-1-4982-8636-7
HARDCOVER ISBN: 978-1-4982-8638-1

Manufactured in the U.S.A.

Figures 2-1, 2-2, 2-3 © Google Earth.
Figures 2-4, 2-6 © NASA.
Figure 2-5 © Fotolia.
Figure 2-7 © Palmer International Partnership LLP.
Figure 5-1 © Christopher Sewell.
Figure 5-2 © Adriana Lalegname.
Figure 6-1 © Mary Ann Sniff.
Figure 6-2 © Granville Sewell.

Contents

Preface

THIS BOOK LOOKS AT a series of issues that cause Christians to doubt.

The first is the widely-believed idea that science has eliminated the need for a Creator, that it can now explain how we got here without design. This idea is easily countered; chapters 1 and 2 present simple but powerful evidence showing that Darwin's implausible explanation for evolution has become even more implausible in recent years, leaving intelligent design as the only reasonable explanation for the origin and development of life. The arguments in these chapters do not require a strong scientific background to understand; for a more extensive presentation of the evidence, the reader should look at my recent *Discovery Institute Press* book *In the Beginning and Other Essays on Intelligent Design, 2nd edition,*[1] or any of a number of other recent books on intelligent design (ID), such as *Darwin's Black Box,*[2] *The Edge of Evolution,*[3] or *Darwin's Doubt.*[4] Section 2.5 shows that design is now being discovered not only in biology but also in the laws of physics themselves, which seem to be very fine-tuned for life.

After making the case for intelligent design in chapters 1 and 2, in the remaining chapters I attempt to deal with, from a

1. Sewell, *In the Beginning.*
2. Behe, *Darwin's Black Box.*
3. Behe, *The Edge of Evolution.*
4. Meyer, *Darwin's Doubt.*

non-fundamentalist point of view, some of the theological problems many educated people have with the Bible and with Christianity. These problems are more difficult, and since chapters 3-6 are about theology, I am naturally not as sure of my conclusions there as in the first two scientific chapters. Nevertheless, I believe that some of the most important insights into both the scientific and theological problems can be made by applying a little common sense, without relying on ideas that can only be understood by the "experts."

Here is a summary of the theological chapters:

3. *The Light of the World*. Before looking at the problems educated people have with Christianity, I explain why I am a Christian, in spite of these problems.

4. *The Bible*. This chapter looks at some of the problems with the Bible, concluding with C. S. Lewis that "[The Bible] is not 'the Word of God' in the sense that every passage, in itself, gives impeccable science or history. It carries the Word of God; and we receive that word from it not by using it as an encyclopedia or an encyclical but by steeping ourselves in its tone or temper and so learning its overall message."

5. *Is the Gospel Good News or Bad?* Section 5.1 looks at the Christian ideas of resurrection and judgment, both of which are very difficult for modern minds to take seriously. Section 5.2 deals with a doctrine which has been taught in many Christian churches over the centuries, which was particularly repulsive to Darwin himself and which may have been responsible for much of his antagonism toward Christianity. As readers will see, I also find this doctrine unreasonable and unbiblical. The last section looks at the meaning of the cross.

6. *Is God Really Good?* This chapter looks at the "problem of pain"—how can we reconcile the idea of a loving God with the pain we experience in the world God created?

Of course, you do not have to believe anything in chapters 3-6 of this book or anything in the Bible to believe in intelligent design;

pre-Columbian American Indians, for example, had never heard of the Bible, yet most of them believed plants and animals were designed. In fact, some intelligent design advocates are uncomfortable with a book that combines chapters on intelligent design with explicitly Christian chapters because it might encourage those who claim that ID proponents do not understand the difference between science and religion. Most of us do understand the difference; we are just interested in both.

And so are ID critics. In fact, I have been making the case for ID for many years, and my experience has convinced me that most of the angriest critics of intelligent design will never be persuaded by logic or evidence because their opposition to ID is based primarily on *religious* convictions. In a June 15, 2012, post at www.evolutionnews.org, Max Planck Institute biologist W. E. Lönnig said "Normally the better your arguments are, the more people open their minds to your theory, but with ID, the better your arguments are, the more they close their minds, and the angrier they become. This is science upside down."

This has been my experience as well; these angry opponents of ID, at least in the Western world, do *not* really have trouble seeing the obvious evidence for design in the living world. They simply have problems with the Bible, which they see as the primary competition on origins, and find Christianity—as they have been taught it—unattractive, and so they prefer materialist explanations of origins, no matter how implausible. For these people, just presenting the scientific evidence for intelligent design in Nature, as done in chapters 1-2, is not sufficient. They need answers for the more difficult theological questions, which I have attempted to address in the final chapters.

1

What is Intelligent Design?

The following article appeared in the on-line version of Human Events *(www.humanevents.com) on December 16, 2013, and in the* El Paso Times *the previous day.*

THE DEBUT AT #7 on the *New York Times* best seller list of Stephen Meyer's new book, *Darwin's Doubt,*[1] is evidence that the scientific theory of intelligent design (ID) continues to gain momentum. Since critics often misrepresent ID and paint ID advocates as a fanatical fringe group, it is important to understand what intelligent design is and what it is not.

Until Charles Darwin, almost everyone everywhere believed in some form of intelligent design (the majority still do); not just Christians, Jews, and Muslims, but almost every tribesman in every remote corner of the world drew the obvious conclusion from observing animals and plants that there must have been a mind behind the creation of living things. Darwin thought he could explain all of this apparent design through natural selection of random variations. In spite of the fact that there is no direct evidence that natural selection can explain anything other than very minor

1. Meyer, *Darwin's Doubt.*

adaptations, his theory has gained widespread popularity in the scientific world simply because no one can come up with a more plausible theory to explain evolution *other than* intelligent design, which is dismissed by most scientists as "unscientific."

But, in recent years, as scientific research has continually revealed the astonishing dimensions of the complexity of life, especially at the microscopic level, support for Darwin's implausible theory has continued to weaken, and since the publication in 1996 of *Darwin's Black Box*[2] by Lehigh University biochemist Michael Behe, a growing minority of scientists have concluded, with Behe, that there is no possible explanation for the complexity of life other than intelligent design.

But what exactly, do these "ID scientists" believe? There is no general agreement among advocates of intelligent design as to exactly where, when, or how design was manifested in the history of life. Most, but not quite all, accept the standard timeline for the beginning of the universe, of life, and of the major animal groups. (Meyer's book focuses on the sudden appearance of most of the animal phyla in the "Cambrian explosion," some 500 million years ago.) Many, including Michael Behe, accept common descent. Probably all reject natural selection as an adequate explanation for the complexity of life, but so do many other scientists who are not ID proponents. So what exactly do you have to believe to be an ID proponent?

Perhaps the best way to answer this question is to state clearly what you have to believe to *not* believe in intelligent design. Peter Urone in his 2001 physics text *College Physics* writes, "One of the most remarkable simplifications in physics is that only four distinct forces account for all known phenomena."[3] The prevailing view in science today is that physics explains all of chemistry, chemistry explains all of biology, and biology completely explains the human mind; thus physics alone explains the human mind and all it does. This is what you have to believe to not believe in intelligent design: that the origin and evolution of life and the evolution

2. Behe, *Darwin's Black Box.*
3. Urone, *College Physics*, 99.

of human consciousness and intelligence are due *entirely* to a few unintelligent forces of physics. Thus you must believe that a few unintelligent forces of physics alone could have rearranged the fundamental particles of physics into computers and science texts and jet airplanes.

Contrary to popular belief, to be an ID proponent you do not have to believe that all species were created simultaneously a few thousand years ago or that humans are unrelated to earlier primates or that natural selection cannot cause bacteria to develop a resistance to antibiotics. If you believe that a few fundamental, unintelligent forces of physics alone could have rearranged the basic particles of physics into Apple iPhones, you are probably not an ID proponent, even if you believe in God. But if you believe there must have been more than unintelligent forces at work somewhere, somehow, in the whole process— congratulations, you are one of us after all!

2

The Case for Intelligent Design

2.1 WHY EVOLUTION IS DIFFERENT

IN THE CURRENT DEBATE between Darwinism and intelligent design, the strongest argument made by Darwinists is this: in every other field of science, naturalism has been spectacularly successful; so why should evolutionary biology be so different?

Joseph Le Conte, professor of Geology and Natural History at the University of California, and (later) president of the Geological Society of America, provides an insight in his 1888 book *Evolution* into the way most scientists think about evolution. In reviewing the fossil record, he writes: "Species seem to come in suddenly, with all their specific characters perfect, remain substantially unchanged as long as they last, and then die out and are replaced by others. Certainly this looks much like immutability of specific[1] forms, and supernaturalism of specific origin."[2] Then in discussing the role of natural selection, he says "Neither can it explain the first steps of advance toward usefulness. An organ must be already useful before natural selection can take hold of it to improve on it."[3]

---

1. "of species"
2. Le Conte, *Evolution*, 251—2.
3. Ibid., 270.

After acknowledging that the fossil record does not support the idea of gradual change and that natural selection can explain everything except anything new, Le Conte nevertheless concludes:

> We are confident that evolution is absolutely certain—not evolution as a special theory—Lamarckian, Darwinian, Spencerian . . . but evolution as a law of derivation of forms from previous forms. In this sense it is not only certain, it is *axiomatic* . . . . The origins of new phenomena are often obscure, even inexplicable, but we never think to doubt that they have a natural cause; for so to doubt is to doubt the validity of reason, and the rational constitution of Nature.[4]

Even most scientists who doubt the Darwinist explanation for evolution are confident that science will eventually come up with a more plausible explanation. That's the way science works; if one theory fails, we look for another one. Why should evolution be so different? Many people believe that intelligent design advocates just don't understand how science works and are motivated entirely by religious beliefs.

Well, perhaps the following story will help critics of intelligent design to understand why evolution *is* different.

Figure 2-1. Moore before first tornado

4. Ibid., 65—6.

Figure 2-2. Moore after first tornado

Here is a set of pictures of a neighborhood in Moore, Oklahoma. The first was taken before the May 20, 2013, tornado hit, and the second was taken right after the tornado.

Fortunately, another tornado hit Moore a few days later and turned all this rubble back into houses and cars, as seen in the third picture below.

Figure 2-3. Moore after second tornado

If I asked you why you don't believe my story about the second tornado, you might say this tornado seems to violate the more general statements of the second law of thermodynamics, such as "In an isolated system, the direction of spontaneous change is from order to disorder."[5] To this I could reply: Moore is not an isolated system because tornados receive their energy from the sun, and the increase in order in Moore caused by the second tornado is easily compensated by decreases outside this open system. Or I might argue that it is too hard to quantify the decrease in "entropy" (disorder) caused by the second tornado, or I could say I simply don't accept the more general statements of the second law of thermodynamics, which should only be applied to thermodynamics, and this tornado does not violate the second law as it applies to *thermal* entropy.

Nevertheless, suppose I further said that I have a scientific theory that explains how certain rare types of tornados, under just the right conditions, really can turn rubble into houses and cars. You doubt my theory? You haven't even heard it yet! If my theory had been studied by the top meteorologists in the world and all agreed that it was plausible, would you take it seriously then? Still no?

Figure 2-4. Earth-like planet soon after it formed

5. Ford, *Classical and Modern Physics*, 619.

Now I have three more pictures for you and two more stories. The first picture shows a certain Earth-like planet in a certain solar system as it looked about four billion years ago. The second shows a large city at the same location about 10,000 years ago. At its prime, this city had tall buildings full of intelligent beings, computers, TV sets, and cell phones inside. It had libraries full of science texts and novels, and airports with jet airplanes taking off and landing.

Figure 2-5. Planet at height of its civilization

Scientists explain how civilization developed on this once-barren planet as follows: about four billion years ago a collection of atoms formed by pure chance that was able to duplicate itself, and these complex collections of atoms were able to somehow preserve their complex structures and pass them along to their descendants, generation after generation. Over a long period of time, the accumulation of duplication errors resulted in more and more elaborate collections of atoms, and eventually something called "intelligence" allowed some of these collections of atoms to design buildings and computers and TV sets and to write encyclopedias and science texts.

Sadly, a few years after the second picture was taken, this planet was hit by a massive solar flare from its sun, and all the intelligent beings died, their bodies decayed, and their cells decomposed into simple organic and inorganic compounds. Most of the buildings collapsed immediately into rubble; those that didn't crumbled eventually. Most of the computers and TV sets inside were smashed into scrap metal; even those that weren't gradually turned into piles of rust. Most of the books in the libraries burned up, the rest rotted over time, and you can see the final result many years later in the third picture below.

Figure 2-6. Planet today

This time the second story is natural and believable, it is the first story that is much more difficult to believe. The development of civilization on this planet and the tornado that turned rubble into houses and cars: each *seems* to violate the more general statements of the second law in a spectacular way. Various reasons why the development of civilization does not violate the second law have been given, but all of them can equally well be used to argue that the second tornado did not violate it either. That is, all *except one*: there is a theory which is widely accepted in the scientific world as to how civilizations can develop on barren planets, while there is no widely-believed theory as to how tornados could turn rubble into houses and cars.

Anyone who claims to have a scientific explanation for how unintelligent agents like tornados might be able to turn rubble into houses and cars would be expected to produce some powerful evidence if they want their theory to be taken seriously. The burden of proof should be equally heavy on those who claim to have a scientific explanation for how a few unintelligent forces of physics alone could rearrange the basic particles of physics into computers and encyclopedias and Apple iPhones—and there is no evidence that natural selection of random mutations can explain anything other than very minor adaptations.

My question to those who treat evolution as just another scientific problem is this: can you now at least understand why some of us feel that evolution is a fundamentally different and much more difficult problem than others solved by science and requires a fundamentally different type of explanation?

For a more "scientific" version of this story, see my 2013 BIO-Complexity article "Entropy and Evolution,"[6] which shows why the fact that the Earth is an open system does not mean, as is commonly argued, that atoms can spontaneously rearrange themselves into computers and jet airplanes here without violating the second law, as long as these increases in order are compensated by even greater decreases outside our open system (so that the total "order" in the universe, or any isolated system containing the Earth, still decreases). In fact, the entropy change equations upon which this widely-used "compensation" argument is based actually support, on closer examination, the common sense conclusion that "if an increase in order is extremely improbable when a system is isolated, it is still extremely improbable when the system is open, unless something is entering which makes it not extremely improbable."[7] The fact that order can increase in an open system does not mean that tornados can turn rubble into houses and cars, and it does not mean that computers can appear on a barren planet as long as the planet receives solar energy; something must be entering our open system which makes

---

6. Sewell, *Entropy and Evolution.*

7. Sewell, *Entropy, Evolution and Open Systems*, 170–5.

the appearance of computers not *extremely improbable; for example: computers.*

2.2 HUMAN CONSCIOUSNESS

In my opinion, human consciousness is the biggest problem of all for Darwinism, but since it is hard to say anything "scientific" about consciousness, it is seldom brought up in the debate over origins. Evolutionary biologists talk about human evolution as though they were outside observers and never seem to wonder how they got inside one of the animals they are studying.

Nevertheless, one way to appreciate the problem it poses for Darwinism or any other mechanical theory of evolution is to ask the question: is it possible that computers might someday experience consciousness? If you believe that a mechanical process such as natural selection could have produced consciousness once, it seems you *can't* say it could never happen again, and it might happen faster now, with intelligent designers helping this time. In fact, most Darwinists do believe it could and will happen—not because they have a higher opinion of computers than I do: everyone knows that in their most impressive displays of "intelligence," computers are just doing *exactly* what they are told to do, nothing more or less. They believe it will happen because they have a lower opinion of humans; they simply dumb down the definition of consciousness, and say that if a computer can pass a "Turing test" and fool a human at the keyboard in the next room into thinking he is chatting with another human, then the computer has to be considered to be intelligent, and conscious. With the right software, my laptop may already be able to pass a Turing test and convince me that I am instant messaging another human. If I type in "My cat died last week" and the computer responds, "I am saddened by the death of your cat," I'm pretty gullible, that might convince me that I'm talking to another human. But if I look at the software, I might find something like this:

if (verb == 'died')

```
fprintf(1,'I am saddened by the death of your %s',noun)
end
```

I'm pretty sure there is more to human consciousness than this, and even if my laptop answers all my questions intelligently, I will still doubt there is "someone" inside my Intel processor who experiences the same consciousness that I do, and who is really saddened by the death of my cat, although I admit I can't prove that there isn't.

I really don't know how to argue "scientifically" with people who believe computers could be conscious. About all I can say is: what about typewriters? Typewriters also do exactly what they are programmed by humans to do and have produced some magnificent works of literature. Do you believe that typewriters can also be conscious?

And if you *don't* believe that intelligent engineers could ever cause machines to attain consciousness, why would you believe that random mutations could accomplish this?

2.3 WHY SIMILARITIES DO NOT PROVE THE ABSENCE OF DESIGN

Since the idea that the "survival of the fittest" could produce all the magnificent species on Earth and human brains and human consciousness is so unreasonable, how did such an idea ever become so widely-accepted in the scientific world? There are two reasons.

First, science has been so successful explaining other phenomena in Nature that—understandably—today's scientist has come to expect that nothing can escape the explanatory power of his science. And Darwinism, as far-fetched as it is, is the best "scientific" theory he can come up with for evolution. As microbiologist Rene Dubos put it in *The Torch of Life* "[Darwinism's] real strength is that however implausible it may appear to its opponents, they do not have a more plausible one to offer in its place."[8]

---

8. Dubos, *The Torch of Life*, 59.

But we have already seen in section 2.1 why evolution is a very different and much more difficult problem than others solved by science and why it requires a very different type of explanation.

Second, for most modern minds, the similarities between species not only prove common descent, they prove that evolution was the result of entirely natural causes, even in the absence of any evidence that natural selection can explain the major steps of evolution. The argument is basically, "This doesn't look like the way God would have created things," an argument used frequently by Darwin in *Origin of Species*. But if the history of life does not give the appearance of creation by magic wand, it does look very much like the way we humans create things, through testing and improvements.

In fact, the fossil record does not even support the idea that new organs and new systems of organs arose gradually; new orders, classes and phyla consistently appear suddenly. For example, Harvard paleontologist George Gaylord Simpson writes:

> It is a feature of the known fossil record that most taxa appear abruptly. They are not, as a rule, led up to by a sequence of almost imperceptibly changing forerunners such as Darwin believed should be usual in evolution . . . This phenomenon becomes more universal and more intense as the hierarchy of categories is ascended. Gaps among known species are sporadic and often small. Gaps among known orders, classes, and phyla are systematic and almost always large. These peculiarities of the record pose one of the most important theoretical problems in the whole history of life: Is the sudden appearance of higher categories a phenomenon of evolution or of the record only, due to sampling bias and other inadequacies?[9]

Actually, if we *did* see the gradual development of new orders, classes and phyla, that would be as difficult to explain using natural selection as their sudden appearance. How could natural selection guide the development of the new organs and entire new

---

9. Simpson, *The History of Life*, 149.

systems of interdependent organs, which gave rise to new orders, classes and phyla, through their initial useless stages, during which they provide no selective advantage? French biologist Jean Rostand, in *A Biologist's View*, wrote,

> It does not seem strictly impossible that mutations should have introduced into the animal kingdom the differences which exist between one species and the next . . . hence it is very tempting to lay also at their door the differences between classes, families and orders, and, in short, the whole of evolution. But it is obvious that such an extrapolation involves the gratuitous attribution to the mutations of the past of a magnitude and power of innovation much greater than is shown by those of today.[10]

Rostand says, nevertheless, "However obscure the causes of evolution appear to me to be, I do not doubt for a moment that they are entirely natural."[11]

We see this same pattern, of large gaps where major new features appear, in the history of human technology (and in software development, as discussed in my *Mathematical Intelligencer* article "A Mathematician's View of Evolution"[12]). For example, if some future paleontologist were to unearth two species of Volkswagens, he might find it plausible that one evolved gradually from the other. He might find the lack of gradual transitions between automobile families more problematic, for example, in the transition from mechanical to hydraulic brake systems, or from manual to automatic transmissions, or from steam engines to internal combustion engines. But if he thought about what gradual transitions would look like, he would understand why they didn't exist: there is no way to transition gradually from a steam engine to an internal combustion engine, for example, without the development of new, but not yet useful, features. He would be even more puzzled by the huge differences between the bicycle and motor vehicle

10. Rostand, *A Biologist's View*, 15.

11. Ibid., 18.

12. Sewell, *A Mathematician's View of Evolution.*

phyla or between the boat and airplane phyla. But heaven help us when he uncovers motorcycles and hovercraft. The discovery of these "missing links" would be hailed in all our newspapers as final proof that all forms of transportation arose gradually from a common ancestor, without design.

The evolution of cars

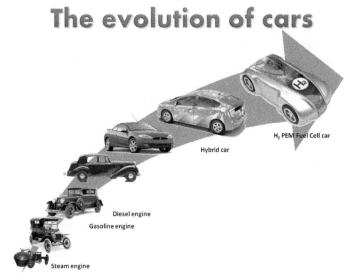

H₂ PEM Fuel Cell car

Hybrid car

Diesel engine

Gasoline engine

Steam engine

Figure 2-7

The similarities between the history of life and the history of technology go even deeper. Although the similarities between species in the same branch of the evolutionary "tree" may suggest common descent, similarities (even genetic similarities) also frequently arise independently in distant branches, where they cannot be explained by common descent. For example, in their *Nature Encyclopedia of Life Sciences* article on carnivorous plants, Wolf-Ekkehard Lönnig and Heinz-Albert Becker note that

> . . . carnivory in plants must have arisen several times independently of each other . . . the pitchers might have arisen seven times separately, adhesive traps at least four times, snap traps two times and suction traps possibly also two times . . . . The independent origin of complex synorganized structures, which are often anatomically

and physiologically very similar to each other, appears to be intrinsically unlikely to many authors so that they have tried to avoid the hypothesis of convergence as far as possible.[13]

"Convergence" suggests common design rather than common descent. The probability of similar designs arising independently through random processes is very small, but a designer could, of course, take a good design and apply it several times in different places to unrelated species. Convergence is a phenomenon often seen in the development of human technology; for example, Ford automobiles and Boeing jets may simultaneously evolve similar new GPS systems.

So if the history of life looks like the way humans, the only other known intelligent beings in the universe, design things— through careful planning, testing, and improvements—why is that an argument *against* design? Somehow we got the idea that God doesn't need to get involved in the details, so he should be able to create anything from scratch, using a magic wand. But no matter how intelligent a designer is, he still has to get involved in the details. That's what design is!

2.4 A *NEW YORK TIMES* ARTICLE

The Darwinist explanation for the development of life is so implausible that the layman would never take such an idea seriously unless he were constantly told that *all* serious scientists believe it; then he begins to think, "Maybe they know something I don't know." But, in fact, many good scientists doubt Darwinism (more and more, actually). To support this claim, I offer here a portion of a November 5, 1980, *New York Times* News Service report:[14]

Biology's understanding of how evolution works, which has long postulated a gradual process of Darwinian

13. Lönnig and Becker, *Carnivorous Plants*, 5.

14. See the rest of this article at www.evolutionnews.org, February 24, 2014.

natural selection acting on genetic mutations, is un-
dergoing its broadest and deepest revolution in nearly
50 years. At the heart of the revolution is something
that might seem a paradox. Recent discoveries have
only strengthened Darwin's epochal conclusion that all
forms of life evolved from a common ancestor. Genetic
analysis, for example, has shown that every organism
is governed by the same genetic code controlling the
same biochemical processes. At the same time, however,
many studies suggest that the origin of species was not
the way Darwin suggested . . . . Exactly how evolution
happened is now a matter of great controversy among
biologists. Although the debate has been under way for
several years, it reached a crescendo last month, as some
150 scientists specializing in evolutionary studies met
for four days in Chicago's Field Museum of Natural History to thrash out a variety of new hypotheses that are
challenging older ideas. The meeting, which was closed
to all but a few observers, included nearly all the leading evolutionists in paleontology, population genetics,
taxonomy and related fields. No clear resolution of the
controversies was in sight. This fact has often been ex-
ploited by religious fundamentalists who misunderstood
it to suggest weakness in the fact of evolution rather than
the perceived mechanism. Actually, it reflects signifi-
cant progress toward a much deeper understanding of
the history of life on Earth. At issue during the Chicago
meeting was macroevolution, a term that is itself a matter of debate but which generally refers to the evolution
of major differences, such as those separating species or
larger classifications . . . . Darwin suggested that such
major products of evolution were the results of very long
periods of gradual natural selection, the mechanism that
is widely accepted today as accounting for minor adaptations . . . . Darwin, however, knew he was on shaky
ground in extending natural selection to account for differences between major groups of organisms. The fossil
record of his day showed no gradual transitions between
such groups, but he suggested that further fossil discoveries would fill the missing links. 'The pattern that we
were told to find for the last 120 years does not exist,'

declared Niles Eldridge, a paleontologist from the American Museum of Natural History in New York. Eldridge reminded the meeting of what many fossil hunters have recognized as they trace the history of a species through successive layers of ancient sediments. Species simply appear at a given point in geologic time, persist largely unchanged for a few million years and then disappear. There are very few examples—some say none—of one species shading gradually into another.

According to this writer, if Darwin's theory on the causes of evolution has to be abandoned, this does not suggest weakness in the "fact" of evolution, only in the "perceived mechanism." When one theory on tectonic plate movements proves wrong, we just propose another; alternative theories on how circulatory, reproductive, digestive, and nervous systems, and hearts and eyes and brains and consciousness all arose on a rocky planet, without design, should be easy to come up with as well!

2.5 THE "FINE-TUNING" OF THE LAWS OF PHYSICS

Le Conte's "axiom" (section 2.1) that everything in Nature must have a natural explanation is a cornerstone of modern thought. Even most scientists who admit that they do not understand how life originated or evolved are still confident that science will eventually explain the origin and development of life in terms of the fundamental, unintelligent laws of Nature alone.

But even if you insist, for philosophical reasons, that it must be possible to explain everything in terms of the basic laws of Nature, it is a well-documented fact that these laws themselves are highly "fine-tuned" to make life possible.

For example, Columbia University astronomer Robert Jastrow (quoted in *The Intellectuals Speak Out About God*) describes what he calls "the most theistic result ever to come out of science":

> According to the picture of the evolution of the universe developed by the astronomer and his fellow scientists,

the smallest change in any of the circumstances of the natural world, such as the relative strengths of the forces of Nature, or the properties of the elementary particles, would have led to a universe in which there could be no life and no man.[15]

As an example, Jastrow cites the forces binding the nuclei of atoms together. If the nuclear force were increased in strength by a small amount, he says, this attraction would have been sufficient to cause all hydrogen nuclei (protons) to fuse together into helium during the early stages of the universe, and there would be no hydrogen left to fuel the stars. On the other hand, if the nuclear force were slightly decreased in strength, the attraction would have been insufficient to drive the nuclear fusion reactions which created elements heavier than helium (such as carbon and oxygen), and it is impossible to imagine how any complex life forms could be constructed out of hydrogen and helium alone.

Jastrow continues,

> It is possible to make the same argument about changes in the strengths of the electromagnetic force, the force of gravity, or any other constants of the material universe, and so come to the conclusion that in a slightly changed universe there could be no life, and no man. Thus according to the physicist and the astronomer, it appears that the universe was constructed within very narrow limits, in such a way that man could dwell in it. This result is called the anthropic principle.

> Some scientists suggest, in an effort to avoid a theistic or teleological implication in their findings, that there must be an infinite number of universes, representing all possible combinations of basic forces and conditions, and that our universe is one of an infinitely small fraction, in this great plenitude of universes, in which life exists.

Now the Darwinist might argue that a different universe, which might be hostile to life as we know it, would simply have

---

15. Varghese, *The Intellectuals Speak Out About God*, 21.

resulted in life forms which are adapted to different conditions. However, we are not talking about conditions which are hostile to life as we know it on Earth, but rather conditions so hostile that any imaginable form of life would be impossible. In *The Problems of Physics*, A. J. Leggett lists several ways in which the development of life depends sensitively on the values of the universal constants and says,

> The list could be multiplied endlessly, and it is easy to draw the conclusion that for any kind of conscious beings to exist at all, the basic constants of Nature have to be exactly what they are, or at least extremely close to it. The anthropic principle then turns this statement around and says, in effect, that the reason the fundamental constants have the values they do is because otherwise we would not be here to wonder about them.[16]

Physicist Steven Hawking discusses some of these fundamental constants of Nature and says, "The remarkable fact is that the values of these numbers seem to have been very finely adjusted to make possible the development of life."[17]

In *Cosmology*, Edward Harrison mentions some other bad outcomes if certain constants were tampered with:

> We first notice that alterations in the known values of c [speed of light], h [Planck's constant], and e [electronic charge] cause huge changes in the structure of atoms and atomic nuclei. Even when the changes are only slight, most atomic nuclei are unstable and cannot exist . . . . We also find that slight changes in the values of c, G [gravitational constant], h, e, and the masses of subatomic particles cause huge changes in the structure and evolution of stars. The majority of universes will actually not contain any stars at all, and in the few that do, the stars either are nonluminous or are so luminous that their lifetimes are too short for biological evolution . . . . Our universe

---

16. Leggett, *The Problems of Physics*, 145—6.
17. Hawking, *A Brief History of Time*, 129.

is therefore finely tuned, and we would not exist if the constants of Nature had different values.[18]

Scientists modeling the big bang have discovered that a universe capable of supporting life requires not only finely-tuned laws, but also initial conditions which are astronomically improbable. Paul Davies in *Other Worlds* appeals to the anthropic principle no fewer than ten times to explain benevolent features of our universe. Citing the calculations of various physicists and astronomers, he notes that fine-tuning of various laws is required (e.g., the strengths of the strong and weak nuclear forces must be just right), but he also shows that, for example, if the matter in the early universe were distributed a tiny bit more—or less—uniformly, or if the material density were a tiny bit higher—or lower, then the resulting universe would have been very hostile toward the conception and development of any form of life. Davies estimates the odds against one of these coincidences to be 10 to the power 1000000000000000000000000000000 to one. And he adds that "there are probably many more features of the world that are vital to the existence of life and which contribute to the general impression of the improbability of the observed world."[19]

Although Davies recognizes that some may see design in the fortuitous features of our universe, he attempts to defend the multiple universes theory. "If we believe that there are countless other universes, either in space or time, or in superspace, there is no longer anything astounding about the enormous degree of cosmic organization that we observe. We have selected it by our very existence. The world is just an accident that was bound to happen sooner or later," he says. Davies compares the anthropic principle's explanation of why the laws, particles, and forces of physics are so friendly toward life to the traditional scientific explanation of why conditions on Earth are so ideally suited for life: "The many universes theory does provide an explanation for why many things around us are the way they are. Just as we can explain why we are

18. Harrison, *Cosmology*, 111.
19. Davies, *Other Worlds*, 178.

living on a planet near a stable star by pointing out that only in such locations can life form, so we can perhaps explain many of the more general features of the universe by this anthropic selection process."[20]

As Michael Behe points out in *The Edge of Evolution*,[21] however, anthropic selection claims only to explain why we live in a universe which can support intelligent life, not why we live in such a "lush" universe, where the fundamental laws of physics not only make life possible, but also make it interesting. For example, some of the heavier chemical elements (such as copper or uranium), which are probably not vital for life itself, have played a critical role in the progress of science and technology, and the existence and useful chemical properties of these elements can also be traced to the fine-tuning of our physical laws.

But we have to ask ourselves not only why do the gravitational, nuclear, and electromagnetic forces have the strengths that they have and why do electrons, protons, and neutrons have the masses and charges they do, but also why are there particles at all and why are there forces between them? We need to wonder not only why the speed of light is 299,792 km/sec, but also why are there photons?

And we should wonder not only why Planck's constant, which appears in the Schrödinger equation, has such a lucky value, but also why are the motions of all particles governed by this partial differential equation? One of the most surprising things about our universe is the beautiful way in which mathematical equations can be used to model physical processes so elegantly. In the case of macroscopic processes, such as diffusion or fluid flow, we can derive the equations from more basic processes, so that in these cases we feel we "understand" why the mathematics fits the physics. But when we get down to the most fundamental particles and forces, we find they *still* obey an elegant mathematical equation, and we have absolutely no idea why—they just do. There is no conceivable reason why the effect that the fundamental forces have on the

20. Ibid., 145.

21. Behe, *The Edge of Evolution*, 223.

fundamental particles should be given by the solution to a complex partial differential equation like this, except that it results in elements and chemical compounds with extremely rich and useful chemical properties and gives partial differential equation software developers like me some very interesting applications to solve.[22] If the elementary particles interacted by bouncing off each other like tiny balls obeying classical Newtonian laws, chemistry would be dead. In *Partial Differential Equations*, Walter Strauss writes,

> Schrödinger's equation is most easily regarded simply as an axiom that leads to the correct physical conclusions, rather than as an equation that can be derived from simpler principles . . . . In principle, elaborations of it explain the structure of all atoms and molecules and so all of chemistry![23]

Are we to assume that in all these other universes there are still nuclear and electromagnetic forces, electrons, protons, and neutrons, and the behavior of the particles is still governed by the Schrödinger partial differential equation; but the forces, masses, charges, and Planck's constant have different values, generated by some cosmic random number generator? Or perhaps the behavior of particles is governed by random types of partial differential equations in different universes, but there are still many universes in which Schrödinger's equation holds, with random values for Planck's constant? No doubt there were some universes which couldn't produce life because their fundamental equation of chemistry looked just like the Schrödinger equation, but with first derivatives in space where there should be second derivatives, or a second derivative in time where there should be a first derivative, or the complex number i was missing, or the linear Vu term was replaced by a nonlinear term Vu^n, where n is not equal to one.[24] The fundamental equation of chemistry appears itself to be finely-tuned.

22. Sewell, *The Numerical Solution of Ordinary and Partial Differential Equations*, 278—80.

23. Strauss, *Partial Differential Equations*, 18.

24. Any of the changes listed—and others not listed—would fundamentally

According to the picture drawn by the popular media, primitive man attributed many phenomena in Nature to design, but science has progressively removed the need for the design hypothesis from these phenomena one by one, and now a group of religious fanatics is trying to make a last stand in biological origins, where things are most difficult to explain. The true picture is very different; in fact, we are discovering that primitive man was *not* wrong in attributing many "natural" phenomena to design. The design just dates back much farther than he imagined—to the origin of the universe. And of course all of the arguments in this chapter take for granted that once the right conditions to support life are present, life can spontaneously develop, an assumption for which there is absolutely no supporting evidence. As noted atheist Richard Dawkins admitted in the movie *Expelled*, no one really has any idea how life could have originated.

It is difficult to argue with those who appeal to "anthropic selection" to explain improbable circumstances; about all you can say is that there is a simpler explanation. But other universes are by definition beyond observation, so that the anthropic principle is untestable and therefore unscientific. It is interesting to see how those who for many years have criticized the creationists for inventing an agent external to our universe to account for the appearance of man are now reduced to inventing other universes to explain our existence.

Fred Heeren[25] illustrates the silliness of the idea that, given enough universes, everything will eventually happen. If there are enough universes, he says, one of them would be just like ours except that in that one Elvis Presley kicked his drug habit, got involved in Tennessee politics, and became president of the United States. It seems much simpler to believe that our universe *appears* to be cleverly designed because it *is* cleverly designed.

---

alter the nature of the solutions, and chemistry as we know it would not exist.

25. Heeren, *Show Me God*.

3

The Light of the World

I BELIEVE THAT MY faith in God rests on a very solid foundation of reason. It is hard to imagine anything more unreasonable than the idea that the universe as we know it, with its marvelous laws of physics and mathematics and the magnificent forms of life which are to be found on our Earth, could have arisen without intelligent design.

My Christian beliefs, on the other hand, are not backed up by nearly as much reason or logic. There are some logical and historical reasons for believing that Jesus was no ordinary man: for example, the fact that someone who never commanded an army or held any political office could command such a following for so many centuries. But I believe primarily because it is my experience that the teachings of Jesus continually illuminate and make sense of life and that life works much better when we follow his teachings. Jesus often used the image of light illuminating darkness to describe his purpose and mission on Earth. "I am the light of the world. Whoever follows me will never walk in darkness but will have the light of life," he said. God has shown us, through the teachings and example of Jesus, how the human life is intended to be lived.

Jesus taught, "Do to others what you would have them do to you, for this sums up the law and the prophets," and "Love the Lord . . . and love your neighbor as yourself. All the law and the prophets hang on these two commandments," and "My command is this: Love each other as I have loved you." He taught us to love our neighbor, and by suffering with us voluntarily, even to the extent of submitting to a cruel death on the cross, he gave us the supreme example of love.

He said "Love your enemies, and pray for those who persecute you," and he healed the severed ear of one of those who had come to take him to trial. He said "If someone strikes you on the right cheek, turn to him the other also," and when they put on him a crown of thorns and a purple robe, mocked him, saying "Hail, King of the Jews," and struck him, he said nothing. He taught, "Blessed are the meek," and in his death he fulfilled the prophecy of Isaiah, "as a sheep before her shearers is silent, so he did not open his mouth."

When he saw how the guests at a wedding party fought over the seats of honor, he said, "Everyone who exalts himself will be humbled, and he who humbles himself will be exalted." And he told the story of a Pharisee and a tax-collector who went to the temple to pray: "The Pharisee stood up and prayed about himself, 'God, I thank you that I am not like all other men . . . .' But the tax-collector . . . beat his breast and said, 'God, have mercy on me, a sinner.' I tell you that this man, rather than the other, went home justified before God." When two of his disciples argued about who was to be the greatest in the kingdom, Jesus said, "Whoever wants to become great among you must be your servant, and whoever wants to be first must be your slave." He told us to be humble before others, and the man from whose birth the world counts time knelt down and washed his disciples' feet, wiping them with the towel he was wearing.

Jesus warned, "Do not store up for yourselves treasures on Earth," and "Be on your guard against all kinds of greed; a man's life does not consist in the abundance of his possessions," and "You cannot serve both God and money." He told the story of "a rich man, who was dressed in purple and fine linen and lived in luxury every day. At his gate was laid a beggar named Lazarus, covered with sores, and longing to eat what fell from the rich man's table." The parable ended with the rich man in torment and the poor man in Paradise. He taught us not to put our trust in riches, and—in contrast to many other religious leaders—he lived his entire life in poverty with "no place to lay his head" at times, even though his many followers would no doubt have gladly given him all he needed for a life of comfort.

He taught, "Do not judge, or you too will be judged," and when the Pharisees and scribes complained that "this man welcomes sinners and eats with them," he told a story of a shepherd who lost one sheep and left the other ninety-nine to hunt for it. Likewise, he said, "There is more rejoicing in heaven over one sinner who repents than over ninety-nine righteous persons who do not need to repent." Jesus continued, telling the famous story of the prodigal son, who squandered all his money on wine, women, and song in a distant land and returned when he was broke and hungry. The father, who represents God, welcomed the errant son back with open arms and explained to his complaining older brother, "We had to celebrate and be glad, because this brother of yours was dead, and is alive again; he was lost and is found." To those who criticized him for associating with tax-collectors, prostitutes, and other sinners, he replied, "I have not come to call the righteous, but sinners to repentance." He taught us to remove the log from our own eye before we are critical of the speck in our brother's eye, and when a woman caught in adultery was brought before him, he told those who wanted to stone her, "If any one of you is without sin, let him be the first to throw a stone at her." Then he said to her, "Go now and leave your life of sin."

When a lawyer, referring to the command to "love your neighbor" asked who his neighbor was, Jesus explained through a parable that his neighbors included the despised Samaritan race. He taught us that all people are equal in God's eyes, and later he astonished his disciples by talking to and drinking from a well with a Samaritan woman—an unthinkable act for a Jewish man.

Jesus said that in the judgment God will say to the righteous, "Come . . . for I was hungry and you gave me something to eat, I was thirsty and you gave me something to drink, I was a stranger and you invited me in, I needed clothes and you clothed me, I was sick and you looked after me, I was in prison and you came to visit me . . . . Whatever you did for one of the least of these brothers of mine, you did for me." He told a story of a "Good Samaritan" who stopped to help a man who had been beaten and robbed, after a priest and a Levite had passed him by. "Go and do likewise," he told his listeners. "When you give a luncheon or dinner, do not invite your friends . . . or your rich neighbors," he taught. "If you do, they may invite you back and so you will be repaid. But when you give a banquet, invite the poor, the crippled, the lame, the blind, and you will be blessed."

Jesus repeatedly emphasized that what is in a man's heart, not superficial religious rituals, makes him right with God. The Pharisees criticized his disciples for not following their hand-washing ritual before eating, saying that they were defiled by not doing so. Jesus replied, "Out of the heart come evil thoughts, murder, adultery, sexual immorality, theft, false witness, slander. These are what make a man unclean, but eating with unwashed hands does not make him unclean." He rebuked the religious leaders, saying, "you devour widows' houses, and for a show make lengthy prayers . . . on the outside you appear to people as righteous, but inside you are full of hypocrisy and wickedness." When a Samaritan woman asked Jesus whether God should be worshiped on a certain mountain in Samaria or in Jerusalem, Jesus said that *where* wasn't important, but "a time is coming and has now come when the true worshipers will worship the Father in spirit and in truth." He quoted Hosea when his disciples, picking corn, were criticized for violating the Sabbath law—a law which was designed simply to ensure that workers were given a day of rest each week but which the Pharisees had turned into a complicated set of religious rules: "I desire mercy, not sacrifice, and acknowledgment of God rather than burnt offerings." He further cited an Old Testament story in which David and his comrades ate the "holy" bread of the tabernacle, which was to be eaten only by the priests, when there was nothing else to eat, to show that human need takes precedence over religious ceremony. Then he said, "The Sabbath was made for man, not man for the Sabbath"—God gave us his laws, not to make life harder, but to make it better.

Jesus taught us to forgive others, as God forgives us. "How many times shall I forgive my brother when he sins against me? Up to seven times?" Peter asked him. "I tell you, not seven times, but seventy times seven," he answered. He then told a parable, comparing God to a king who forgave one of his servants an extremely large debt. The servant turned around and put a fellow servant in jail for non-payment of a tiny sum, and when the king heard of this, he called the first servant in and said, "You wicked servant. I cancelled all that debt of yours because you begged me to. Shouldn't you have had mercy on your fellow servant just as I had on you?" Jesus taught "Pray for those who persecute you," and when those who hated him nailed him to a cross, mocked him, and spat upon him, he prayed, "Father, forgive them, for they do not know what they are doing."

Jesus especially warned against religious hypocrisy. He reserved his only angry words for the scribes and Pharisees, who "do everything they do for men to see," and his only recorded act of violence was to overturn the tables of the merchants and money-changers who had turned the temple into a "den of thieves."

Jesus brought the light of love into a world dark with hatred. He told us in the Sermon on the Mount that we are to spread this light: "You are the light of the world . . . let your light shine before men, that they may see your good deeds and praise your Father in heaven."

But Jesus also brought the light of hope into a world dark with despair. He assured us that God loves us: "Look at the birds of the air; they do not sow nor reap or store away in barns, and yet your heavenly Father feeds them. Are you not much more valuable than they?" At a time when most of the world regarded God as an indifferent or even tyrannical monarch, Jesus told us we could come before God and say, "Our Father in heaven . . . ." To those weary from pain and heartbreak, he said, "Come to me, all you who are weary and burdened, and I will give you rest."

Who was this man who gave us the Sermon on the Mount, who said, "Do to others what you would have them do to you" and "Everyone who exalts himself will be humbled, and he who humbles himself will be exalted" and "Blessed are the merciful, for they will be shown mercy . . . blessed are the peacemakers" and "Love your enemies and pray for those who persecute you," who said he "came not to be served, but to serve," and who lived a life of service and self-denial?

Like most people, I wonder why God would use a first century Jewish carpenter as his primary spokesman; I wonder why he doesn't speak to us more loudly and more clearly. But when I look at what Jesus did and taught, I also wonder, was this a madman who deluded himself and millions of others into believing he was divine? Or was he really Immanuel—"God with us"?

4

The Bible

4.1 HISTORICAL ACCURACY OF THE BIBLE

MUCH OF THE WORLD seems to gravitate toward either the view that the Bible is just a collection of Hebrew myths or else the opposite view that every word of it was dictated by God himself. What should we think of the Bible? Is it "inspired" by God, and if so, in what sense?

While the earliest portions of the Bible may sound more mythological than historical, much of the Bible is firmly rooted in history. Many of the events, people, and places mentioned in the Bible are also mentioned by secular historians in Israel and in neighboring lands. For example, the Jewish historian Josephus mentions the arrest of John the Baptist by Herod, and Annas and Caiaphas, high priests during the life of Jesus, are mentioned by secular historians. Another example is the biblical story of how Herod slaughtered all the male children under the age of two in the region of Bethlehem for fear that Jesus—the "king of the Jews"—would usurp his throne. This is quite consistent with what is known from extra-biblical sources about Herod the Great's personality, for he is reported to have killed so many of his relatives out of fear of a coup that it was said, "It is safer to be Herod's pig than Herod's son." Then Matthew reports that after Herod the

Great died, Joseph and Mary chose to return from exile in Egypt to Nazareth, rather than Judea, because Archelaus was now ruler in Judea. From other sources, it is known that the new ruler in Galilee—Herod Antipas—was indeed much less violent than Archelaus, who took his father's place in Judea. A good commentary on the New Testament such as William Barclay's *The Daily Study Bible Series,*[1] which contains the information cited above, will document hundreds of other details recorded in the Gospels which are consistent with what is known from other sources about the customs, people, and places of first century Israel.

While many things are known from secular sources about Pilate, the Herods, and other rulers mentioned in the Gospels, Jesus himself is not mentioned by any contemporary historians. This is not surprising, since the life and death of a poor itinerant preacher would not be considered of historical importance at the time. However, Jesus is mentioned by several historians not too long after his death. For example, the Roman historian Tacitus wrote in 112 AD, "Christus [was] executed by the governor Pontius Pilatus when Tiberius held power. The pernicious creed, suppressed at the time, was bursting forth again, not only in Judea, where this evil originated, but even in Rome."[2]

Naturally, as we move back in time in the Old Testament, archeological and historical confirmations of biblical events become fewer and farther between, but even many Old Testament stories have found extra-biblical support. References to the biblical protagonists Jehu, Omri, Hazael, and Ahab, for example, have been found in inscriptions from neighboring countries, and there is archeological evidence for the existence of many of the cities and nations mentioned in the Old Testament as far back as (possibly) Abraham. Any book on biblical archeology will contain abundant data from extra-biblical sources which confirm narratives in the New Testament and later portions of the Old Testament. An example appears in the following paragraph from Cornfeld and Freedman:

1. Barclay, *The Daily Study Bible Series*.
2. Tacitus, *Tacitus Annals*, 325.

The publication in 1963 by D.J.Wiseman of additional tablets of the roughly contemporary "Chronicles of Chaldean Kings" provided a much fuller picture of the course of events in the Near East in the period preceding the destruction of Jerusalem than had previously been available. These clay tablets, inscribed in Babylonian cuneiform, describe the political and military developments in the years 616-595 BC. They tell of the fall of Nineveh, the Assyrian capital, to the Babylonians and Medes in 612 BC and document Babylonian relations with Necho of Egypt. They continue with Nebuchadnezzar's conquest of Syro-Palestine as implied in 2 Kings 24 and list the tribute received from the kings of these lands. Among those giving tribute is Jehoiakim, who remained loyal until the Babylonians were defeated by the Egyptians in 601 BC. Spurred on by the pro-Egyptian false prophets who saw in this Babylonian setback confirmation of their predictions of well-being, the king of Judah rebelled openly against Babylon. The inevitable response came as soon as the Babylonians were able to settle problems elsewhere in the empire. The fifth paragraph of the chronicle relates the capture of Jerusalem in 597 and the deportation of Jehoiachin (the young son and successor of Jehoiakim, who died during the siege). Jehoiachin's uncle, Zedekiah, another son of the great Josiah, served as the last reigning king of the house of David.[3]

Cornfeld and Freedman point out apparent conflicts between the Bible and archeology and do not hesitate to express doubts about many other biblical accounts. Nevertheless, in their introduction they state, "While it is true that, for the most part, archeology has substantiated and illuminated the biblical story, the biblical archeologist must limit his deductive thinking by rigid scientific discipline," and also "In general we may cite the late E. A. Speiser: 'Independent study helps to increase one's respect for the received material beyond the fondest expectations of the confirmed traditionalists.'"[4]

3. Cornfeld and Freedman, *Archaeology of the Bible*, 173—4.
4. Ibid., 2.

4.2 PROBLEMS WITH THE BIBLE

Nevertheless, there are problems. First, there appear to be histori-cal errors. A typical example is brought out by William Barclay in his commentary on Luke 2. In this chapter, Luke states that the census which forced Joseph and Mary to travel to Bethlehem "took place when Quirinius was governor of Syria." Barclay points out that Quirinius became governor of Syria in 6 AD (though he did hold another government post in Syria from 10-7 BC), while the last census before Herod the Great's death was in 8 BC.

Perhaps Luke is as likely to be correct as the secular sources which conflict with his account, and even if it were proved that Luke made a mistake in dating the birth of Jesus, it would not bother most of us, although it might present a problem to the per-son who believes that the Bible was inspired verbally (word for word) by God. Barclay goes on to point out that a written account of a Roman census in Egypt was recently found which indicates that everyone was compelled to return to his city of birth for that census. This discovery lends support to the biblical account that everyone was required by the Roman census to return to his birth city, a point that had been doubted by many historians. Thus it is not impossible that further information might similarly clear up this and other minor conflicts.

Of course, we know that even if the books of the Bible were inerrant in their original form, errors have crept into them during the twenty and more centuries between then and now. With the Dead Sea Scrolls, for example, we are able to compare the copies of Isaiah and some other Old Testament books we have now with earlier copies than were previously available, and we can see that minor changes have been introduced. Neither this nor the minor conflicts with archeology are of much concern to me, or to the majority of Christians. I could live with an imperfect, but basically accurate Bible.

The miracles described in the Bible present a more serious second type of problem. For many people—including many who believe in God—these miracles present a problem because they do not believe God can overrule the laws of Nature. For others, they present a problem because at least some of the biblical accounts of God's interventions in the affairs of men (such the stories of Noah and the ark or Jonah and the big fish) simply sound much more like mythology than history.

I personally have no trouble believing that the God who created this universe with its magnificent natural laws is able to affect the affairs of men in a "supernatural" way, circumventing the natural laws he has himself designed. In fact, as discussed in chapter 8 of *In the Beginning*,[5] with the introduction of quantum mechanics with its "principle of indeterminacy" into modern physics, the distinction between what is natural and what is supernatural is blurred; there is now a supernatural element (forever beyond the ability of science to explain or predict) in all "natural" events. Although things which we would call miracles may be astronomically improbable, modern physics cannot say that anything is impossible. But the history of life on Earth is replete with extremely improbable events and turning a mass of inanimate molecules into zebras, giraffes, and conscious human beings in a few billion years (whether gradually or—as the fossil evidence[6] seems to indicate—through sudden jumps) is not in any real sense less miraculous than turning water into wine in an instant.

Here it is important to remember what we learned in chapters 1 and 2, that God really did create entire new animal classes and phyla at specific times and specific places in the history of life. This idea is rejected, even ridiculed, by the majority of scientists, not because there is any shortage of evidence for it, but only because they reject divine intervention *a priori*, on principle, and prefer naturalistic explanations for the origin of species, no matter how implausible. Why God doesn't seem to work miracles today—at least not openly, for everyone to see—even when we need his help the most is a very difficult question, which I attempt to address in chapter 6. But whether miracles continue to happen today or not, we can be sure they have happened at times in the past; there

5. Sewell, *In the Beginning*.

6. See section 2.4, for example.

is simply no other explanation for what we see today in the living world. That is why I believe it is not reasonable to reject all miracles reported in the Bible *a priori*.

Even atheists now recognize that this universe was created suddenly about 15 billion years ago (see chapter 6 of *In the Beginning*), and since there were no natural causes before Nature came into existence, there is no chance of explaining this "big bang" in terms of natural causes. And I can't see that healing a leper is any more difficult than bringing time, space, matter, and finely-tuned laws of physics[7] into existence out of nothingness, with a big bang. Or consider the idea that the spirit of God could come down and enter into a human body and be born to a virgin; this is something which is hard for our scientifically trained minds to accept because it is so foreign to our experience. But the idea that my own human spirit could become, through the "normal" birth process, so united with a human body that I would call that body "me" is just as incomprehensible. It is less shocking to us only because we see it happen every day, not because we are really any closer to understanding it. We are just more accustomed to some miracles than others.

If you remove all miracles from Christianity, it is just another nice philosophical system. And if I did not believe that Christ really rose from the dead, I would not see any reason to trust the other tenets of Christianity. But some of the early stories in the Bible simply sound more like myths, with perhaps some basis in fact and perhaps some symbolic meaning, than accurate historical accounts of real events. I doubt, for example, that the story of Adam and Eve was ever intended to be taken as more than an allegory. It seems to me that the story about how "mankind" (the literal translation of "Adam") ate of the "tree of the knowledge of good and evil" and became "like God" and had to leave Paradise is not about two historical individuals. Rather, it tells us that sin and sorrow originated when God took the human animal and made him "like God," giving him the ability to think and make decisions on his own. The "tree of the knowledge of good and evil" is the free

7. See section 2.5.

will which God gave us, which brought not only pain and evil into the world, but also joy and goodness. If the writer had intended for us to take this story literally, I believe he would have used a species of tree with a less metaphorical name!

We can now see that the Genesis 1 account of creation gives a reasonable general outline of the events of creation. It correctly states that the universe had a definite beginning (as confirmed by the big bang theory, to the dismay of atheists), and that there was a progression in time from the creation of the stars, Sun, and Earth to the creation of the sea creatures, to the land animals, and finally to the creation of man. Notice that the God of the Bible creates through testing and improvements, like we do. After each creation, God "saw that it was good" and proceeded to improve on his designs, and at the end he "rested from all the work of creating that he had done." We have already noticed in section 2.3 that the fossil record also suggests that God created step-by-step (though not really gradually), like we do, through testing and improvements. This does not mean, by the way, that God did not have us in mind from the beginning: human technology also progresses step-by-step even when the designers have a clear idea from the beginning of the ultimate goal.

Genesis 1 paints a more accurate picture of natural history than other creation stories of its time. However, it is clearly a literary account rather than a scientific one, and it is certainly inaccurate in many details. "Noah and the flood" is another story which is impossible to take as more than an allegory, but is there a message in this story? I think there is: the message is that God was genuinely surprised and disappointed at how badly the animal that he had created "in his image"—i.e., with its own free will— had turned out and almost decided (probably more than once, and perhaps more recently than we think!) to end this experiment in human freedom, but found enough good in mankind to make the experiment worth continuing.

I am not sure which of these very early stories are historically accurate, and I would be less surprised than most people if it were proved that some of them are historical. But some of them just sound so much more like legends than history that I cannot believe they are historical. The early stories in Genesis were not recorded in written form until hundreds of years after the fact, and some of them seem to bring God down to the level of the gods of Greek mythology. The miracles of Jesus, by contrast, are reported by people who were either witnesses (Matthew and John) or interviewed witnesses (Mark and Luke), and they are complete with historical details which are often confirmed by other sources.

Many Christians cling to the "inerrancy" of even the earliest stories of the Bible because they fear that questioning them is only the first step toward abandoning everything in the Bible. They fear that if we examine these stories critically, we will find that the biblical account of creation is only an allegory, and how we really got here can only be discovered by reading *Origin of Species*. But I have found that even many fundamentalist Christians are willing to recognize there are problems with the Bible once they see that while the biblical account of creation is an allegory, what really happened is that God really did create "the heavens and the Earth" in the beginning.

But there is an even more serious third type of problem I have with portions of the Bible. Parts of the Bible, nearly all in the Old Testament, paint an entirely different picture of God than that painted by Jesus in the Gospels. The most extreme example of this is the story of Saul's battle with the Amalekites, in which he was supposedly told by God to destroy all the men, women, children, and animals in Amalek, and in fact he got into trouble for sparing some of them! How can this possibly be the same God as the one compared by Jesus to a loving father in the parable of the prodigal son? The ways of the Lord are higher than my ways, as Isaiah says, but I still find it impossible to believe that the God described by Jesus would really countenance such a massacre. There are parts of the Bible that I simply wish were not there, yet even in the Old Testament there is so much that is good. There is the comfort of the Psalms ("Even though I walk through the valley of the shadow of death, I will fear no evil, for you are with me"); the wisdom of Proverbs ("He who mocks the poor shows contempt for their

Maker; whoever gloats over disaster will not go unpunished"); and the admonitions of the prophets (Micah: "What does the Lord require of you? To act justly and to love mercy and to walk humbly with your God"; Isaiah: "Is not this the kind of fasting I have chosen, to loose the chains of injustice and untie the cords of the yoke, to set the oppressed free and break every yoke? Is it not to share your food with the hungry and to provide the poor wanderer with shelter?"). There are stories about God's providence for Joseph, whose brothers sold him into slavery, yet he refused to retaliate after reaching high office in Egypt, saying "You intended to harm me, but God intended it for good"; about God's forgiveness for David, who is called "a man after God's own heart" after repenting of murdering a man to steal his wife; about Ruth's loyalty to her foreign-born mother-in-law; and about Job's patience in enduring suffering he did nothing to deserve.

4.3 INSPIRATION

When we look at the Bible as a whole, we see much the same picture as when we look at the creation of life on Earth: powerful evidence of God's presence, yet with details that raise doubts. If God created living things, why didn't he do it in a way that would silence all doubters? If God wrote the Bible, why are there errors in it?

I believe that in some real sense the writers of the Bible were inspired by God. Yet it is clear to me that God did not write the Bible, nor did he dictate it word for word to human secretaries, but that it was written by ordinary human beings and therefore necessarily reflects the individual viewpoints and imperfections of these human writers. (Note that even in normal usage, "inspired by" does not mean "written by" or "dictated by," just "influenced by.")

In Romans 11, for example, the apostle Paul can get caught up in dry theological arguments that only another Pharisee could appreciate. Yet in the next chapter he can launch into some of the most "inspiring" practical Christian teachings:

Love must be sincere. Hate what is evil; cling to what is good. Be devoted to one another in brotherly love. Honor one another above yourselves. Never be lacking in zeal, but keep your spiritual fervor, serving the Lord. Be joyful in hope, patient in affliction, faithful in prayer. Share with God's people who are in need. Practice hospitality. Bless those who persecute you; bless and do not curse. Rejoice with those who rejoice; mourn with those who mourn. Live in harmony with one another. Do not be proud, but be willing to associate with people of low position. Do not be conceited. Do not repay anyone evil for evil. Be careful to do what is right in the eyes of everybody. If it is possible, as far as it depends on you, live at peace with everyone. Do not take revenge, my friends, but leave room for God's wrath, for it is written: "It is mine to avenge; I will repay," says the Lord.

Consider, for a moment, the following portion of Psalms 19:

The heavens declare the glory of God; the skies proclaim the work of his hands. Day after day they pour forth speech; night after night they display knowledge. There is no speech or language where their voice is not heard. Their voice goes out into all the Earth, their words to the ends of the world.

In what sense, if any, was this writing inspired by God? Did God simply tell the Psalmist, word for word, what to write down? Was the human being who put this to ink simply the secretary for God? If so, it is rather meaningless, for it is simply God praising God! Doesn't it have more meaning for us if we see this Psalm as the outpouring of a human heart which has learned, through all of his life experiences, to be in awe of the power and glory of God as seen through Nature?

The epistles of Paul and Peter are wonderful aids in our Christian growth, but I doubt that Paul, when he wrote a personal letter to the Corinthians, had any idea that his words would someday be taken to be God's own words (nor did the early church!). Surely the humble apostle Paul, who said of himself, "I do not even deserve to be called an apostle" would not place his words on a level with those of his "master" Jesus. And yet, listen to the words of this remarkable man:

> Command those who are rich in this present world not to be arrogant nor to put their hope in wealth, which is so uncertain, but to put their hope in God, who richly provides us with everything for our enjoyment. Command them to do good, to be rich in good deeds, and to be generous and willing to share. (I Timothy 6)

or

> The God who made the world and everything in it is the Lord of heaven and Earth and does not live in temples built by hands. And he is not served by human hands, as if he needed anything, because he himself gives all men life and breath and everything else . . . .[God made man] that men would seek him and perhaps reach out for him and find him, though he is not far from each one of us. For in him we live and move and have our very being. (Acts 17; Paul, upon seeing an altar to "an unknown god")

or

> Who shall separate us from the love of Christ? Shall trouble or hardship or persecution or famine or nakedness or danger or sword? . . . No, in all these things we are more than conquerors through him who loved us. For I am convinced that neither death nor life . . . nor the present nor the future, nor any powers, neither height nor depth, nor anything else in all creation, will be able to separate us from the love of God that is in Christ Jesus our Lord. (Romans 8)

Clearly Paul was not just acting as a secretary for God when he wrote these words—these words come directly from the heart of Paul. But knowing that a man who had endured imprisonment, beatings, stonings, and loneliness as a reward for his years of service to God could write "[nothing] will be able to separate us from the love of God" gives, I believe, even greater meaning to these words.

The writers of the Bible were ordinary men who witnessed extraordinary events and who were inspired by God in some sense, but I believe that only Jesus speaks for God. If those who wrote the Bible could speak for God, why do we need Jesus' teachings? If Paul was just writing down what God told him to write, in what sense are the teachings of Jesus more important than those of Paul? If, on the other hand, Paul and the other writers of the Bible were expressing their own points of view, it is as unreasonable to expect that the Bible would be errorless and infallible as it is to expect that God could give man a free will and yet remain in complete control of human events.[8]

Jesus himself made a clear distinction between his words and those of the Old Testament writers. He said, "You have heard that it was said, 'Eye for eye and tooth for tooth.' But I tell you . . . if someone strikes you on the right cheek, turn to him the other also." When I read the words of Jesus in the Gospels, I find myself thinking, this is what God would say if he came to Earth; this is what man most needs to hear. There are other parts of the Bible, mostly in the Old Testament, which leave me much less inspired.

But where, exactly, is God in all of this? Well, this is the problem, and it is the same problem we faced in chapter 1 in looking at the origin and development of life. It is hard to tell exactly where God was during the creation of life and of new species, but when we step back and look at the big picture, we can be certain he was there, somewhere.

It seems we face a similar situation in the "inspiration" of the Bible. When we examine how the Bible was written and how it was decided which books should be included in what we today call the Bible, it is difficult to tell exactly where God is. But when we sit back and look at it as a whole, I believe we will conclude that he was there, somewhere. In the Bible, as in Nature, God reveals himself to us in such a way that he is not best seen through the microscope of scientists or the research of historians, but through

8. And the Bible does not paint a picture of a God who is in complete control of human events. The God of the Bible is often surprised and disappointed, and forced to plan B, by human behavior.

the humble eye of faith. "God chose the foolish things of the world to shame the wise," wrote the Apostle Paul (I Corinthians 1:27). The God of the Bible is a God who likes to use ordinary, or seemingly ordinary, means to accomplish extraordinary ends. His own "son" was born in a stable to a carpenter and his wife, died like a common criminal, and sent twelve average, working-class men out to change the world. They say a picture is worth a thousand words, and rather than thousands of words of dry theology, the Bible contains mostly pictures of the lives of real people, ordinary and extraordinary, good and bad.

Many Christians today believe that the church has always regarded the Bible as verbally inspired by God and that any movement away from this belief threatens the foundations of Christianity. But this is not true. Since the very beginning of the church, Christians have wrestled with the question of biblical authority and have come to many different conclusions. Martin Luther said that the books of Esther and Revelation do not belong in the Bible, and, according to Bernhard Lohse:

> Luther . . . did not develop any doctrine of verbal inspiration. [Lohse's footnote: The doctrine of verbal inspiration was developed after the middle of the 16th century.] Quite the opposite is true. For example, Luther was openly critical of the substance of certain portions of Scripture, particularly of the letter of James.[9]

And yet no one is better known for his emphasis on the authority of scripture than Martin Luther, who wrote:

> Neither councils, fathers, nor we, in spite of the greatest and best success possible, will do as well as the Holy Scriptures, that is, as well as God himself has done.[10]

Although it is true that many of those who today subject the Bible to "historical criticism" assume every biblical account to be false until confirmed by other sources (an approach they would never apply to any other historical document), Martin Luther

9. Lohse, *Luther*, 155—6.
10. Ibid., 156.

stands as a witness to the fact that Christian faith and zeal need not hinge on the accuracy of every word in the Bible.

I have to admit that the errors and difficulties in the Bible do disturb me, sometimes quite a lot. But the best antidote for me, for the doubts generated by biblical difficulties, is to sit down with William Barclay's commentaries[11] and read a while. Barclay does not hesitate to point out inaccuracies in the text or to question an opinion of the apostle Paul. Yet one never reads very long without gaining the clear impression that the Bible is a very special book, that these minor problems are of only minor importance, and that, as a whole, the Bible is reliable. And one is left assured that the Bible is, in some real sense, the word of God to us, filtered through the human minds of its authors.

C. S. Lewis, in *Reflections on the Psalms*, says of the Bible:

> The human qualities of the raw materials show through. Naivete, error, contradiction, even wickedness are not removed. The total result is not "the Word of God" in the sense that every passage, in itself, gives impeccable science or history. It carries the Word of God; and we ... receive that word from it not by using it as an encyclopedia or an encyclical but by steeping ourselves in its tone or temper and so learning its overall message.[12]

The Bible is important because it describes the events leading up to the birth of Jesus, it contains an account of the life and teachings of Jesus and it describes the establishment and early history of Christ's church. But we must remember that it is Jesus, not the Bible itself, who stands at the center of the Christian faith.

11. Barclay, *The Daily Study Bible Series*.
12. Lewis, *Reflections on the Psalms*, 111—2.

5

Is the Gospel Good News or Bad?

5.1 THE RESURRECTION

Jesus spoke frequently of a resurrection and of a judgment after death, and one of the central doctrines of Christianity is that Jesus himself rose from the dead, a testimony to our own immortality.

Both resurrection and judgment are ideas which are very difficult for modern minds to take seriously. The idea that I could someday awaken to find myself in a strange and fantastic new world that God has created for me, using powers beyond my comprehension, is an idea that is so bizarre and speculative that I would also have to consider it a fairy tale, except for one thing: it has already happened once. If God is able to create one world and put me inside one body, I don't know why he couldn't do it again, with a new world and a new body. That new world seems so vague and mysterious to me now that it is hard to take seriously; but then, before I was incarnated into a body here I could never have imagined what this strange new world would be like. Yet here I am, in a very real and fantastic—if imperfect—world. I could surely not have ever imagined a world of mountains and rivers, jungles and waterfalls, butterflies, elephants, giraffes and hippopotamuses, and art and music and mathematics.

Figure 5-1. Isaac Explores His Strange New World

As we noted in section 2.3, while God is far more intelligent than we are, he nevertheless seems to create through testing and improvements, like we do.[1] So maybe God is creating another universe for us, with even better-designed laws of Nature and a new world in it, with major improvements over our old one. (Just the fact that Hitler, Stalin, and bin Laden would not be there would be very helpful!) Just speculation, of course.

Much of the appeal of atheism has always been that while you have to believe in things that are so improbable that they are impossible, at least you didn't have to engage in wild speculation like this about another reality outside our observable universe, beyond the reach of science. But, as discussed in section 2.5, to explain the fine-tuning of our laws of Nature, atheists are now forced to imagine an infinite or nearly infinite number of other universes with different laws and conditions. We are only asking you to imagine one!

1. This is also suggested by the Genesis creation story: after each step God "saw that it was good" and proceeded to the next step.

Figure 5-2. A Caterpillar is Resurrected as a Butterfly

The idea that a decomposed, dead body could be replaced by a new body someday, somewhere, seems impossible. But to me it seems equally impossible that an ugly caterpillar could enter a tomb and be resurrected as a beautiful new butterfly, and yet a butterfly with many entirely new organs is constructed out of the dissolved and recycled parts of a caterpillar every day in a chrysalis, as the film *Metamorphosis*[2] documents so magnificently. This film includes photography (through magnetic resonance imaging) of the transformation as it happens within the chrysalis. If you find it impossible to believe in the miracle of resurrection, I urge you to watch this, and you will realize that, again, we are just more used to some miracles than others.

I do not find it as hard to believe in a resurrection as many others do—it is the concept of judgment that poses great difficulties for me. Given that our behavior, and even our beliefs, are

2. www.metamorphosisthefilm.com.

heavily influenced by our environment and our heredity, how could God possibly judge us fairly? In fact, many people today believe that *all* human actions are beyond our control, that heredity and environment determine everything we do. We have known for a long time that a person with an X and a Y chromosome is more prone to violence than a person with two X chromosomes, and that a child who was abused by his parents is more likely to abuse his children. And yet not all males become serial killers, and not all abused children grow up to be abusers. We all understand that we do have some control over our own behavior, even if behavior is influenced by many external factors.

Still, it is hard to imagine how God could take into account all those external factors and judge us only for what we can control. And clearly there can be no judgment for babies who died in infancy or others who never developed a sense of right and wrong. And I cannot believe that punishment could really be "eternal" for crimes committed during our short life spans. We can only wait and trust that God will be fair and merciful.

The Sadducees—the religious liberals of Jesus' day—had similar problems with the ideas of resurrection and judgment. They posed a difficult question for Jesus about a woman who had had seven husbands in life: Which one would be her husband in the afterlife? Jesus responded that their error was in thinking that the next life will be just like this one; in the resurrection, people "will neither marry nor be given in marriage." Our problem also is, I think, that our imagination is limited by what we have experienced in this world.

Since Jesus knew the Sadducees did not believe in a resurrection at all, he went on to argue the main point of their question with them. His "proof" that the dead will be raised was as follows (Luke 20): "Moses showed that the dead rise, for he calls the Lord, 'the God of Abraham, and the God of Isaac, and the God of Jacob.' He is not the God of the dead, but of the living." A strange argument at first glance, but it is really the main reason for believing in the resurrection. If Abraham, Isaac, and Jacob are still dead, and their short relationships with God ended with their deaths, what good is God? That is the point Jesus was making. If there is no resurrection and no judgment and the only justice we will ever see is here on Earth, then we would have to conclude that evil is more powerful than good, for there is certainly no final justice here. Sometimes justice is served here, but more often it is not. As the Apostle Paul wrote, "If only for this life we have hope in Christ, we are to be pitied more than all men." (I Corinthians 15:19)

In Colombia in 1989, a presidential candidate is deeply troubled by the corruption which drug money has brought to his country, and he decides to speak out against the drug traffickers. He receives a death threat, but at every opportunity he continues to call for tougher measures against the drug trade. He receives more death threats, but he continues his campaign, knowing very well that, in Colombia, politicians who take on the drug lords have very short life expectancies. This true story ends when the candidate is gunned down at a campaign stop; the drug lord who ordered his assassination continues to live in luxury in the mountains. Pure good versus pure evil, and evil wins. We don't often think about what happens after we die; it's too mysterious and nebulous, and I don't believe God intended for us to be so preoccupied with the next life that we lose interest in this world; he has put a lot of work into this one, and it is a masterpiece. But if we believe this is the end of the story, then who is going to have the courage to take up the fight against corruption and drug money? And if this really is the end of the story, then what an unjust world God has created!

In chapter 6, "Is God Really Good?" I quote from a letter my wife Melissa wrote to our young children when she knew she had lost her battle with cancer. In that quote she talks about how much she learned about love from others during her illness. In another part of the letter, she says:

> My faith may be battered and bruised. I may not understand why God did not send his angels down to protect me and my family from these terrible times . . . But I know that he is still there and someday I will find answers to my many questions.

She could face death bravely because she really believed that this life was only a part of her walk with God and that a better part still awaited her. I hope she was right—she *must* be right—because if she was wrong, God is *not* really good; at least he was not good to Melissa.

The idea of a judgment after death is terribly difficult for our modern minds to take seriously. But, for me, the idea that there will be no final justice—no reward for generosity, kindness, mercy, and courage, and no punishment for selfishness, betrayal, arrogance, and cruelty—is even harder to accept. That would mean that those who are confident that they will never be punished for their corruption and cruelty will be proved right, while those who believe their unselfishness and sacrifices will someday be recognized are deluding themselves. That would mean politicians are smart to only be concerned about what they do and say when the TV cameras are on, and Jesus was giving bad advice when he said, "When you give to the needy, do not announce it with trumpets, as the hypocrites do in the synagogues and on the streets, to be honored by men. I tell you the truth, they have received their reward in full. But when you give to the needy, do not let your left hand know what your right hand is doing, so that your giving may be in secret. Then your Father, who sees what is done in secret, will reward you." (Matthew 6:2-4)

Is it really possible that I will never meet my Creator, that I will never know any more than I do now about what he is like, or why I am here? Is it really possible that Melissa will never find answers to her questions? Is our God the God of the living or of the dead? I wish I had a more scientific argument, and I certainly understand the doubts others have, especially about the idea of judgment, but this is the best I can do: I believe there will be a resurrection and a judgment because I believe that justice must finally prevail.

5.2 IS THE GOSPEL GOOD NEWS OR BAD?

Most of the Christian ideas which I was taught as a child were attractive to me, and I accepted them readily. But I was taught one doctrine which seemed unfair to me even when I was a child and which has always struck me as completely inconsistent with the others. Indeed, much of the ill-will toward the church from outside can be traced to this doctrine, which is still taught in some Christian churches today (although I think not really believed by most even in these churches). There are a very few passages in the New Testament, most notably John 3:18, "he who does not believe stands condemned," the Great Commission of Jesus in Mark 16, and Jesus' statement that "no one comes to the Father except through me" (John 14:6), which seem to imply, and are sometimes interpreted to mean, that all non-Christians will be condemned at the final judgment. "I am come into the world as a light, so that no one who believes in me should stay in darkness," Jesus proclaimed. But why should a person who does the best he can with what light he has be "condemned" simply because the light of Christ has not shined on him? I don't think it is possible to overstate the damage this idea has done to relations between the church and the outside world. Charles Darwin said, "I can indeed hardly see how anyone ought to wish Christianity to be true; for if so, the plain language of the text seems to show that the men who do not believe, and this would include my father, brother and almost all my best friends, will be everlastingly punished."[3]

This question is related to one of the great controversies of the church: the issue of salvation by "faith or works." On the one hand, many teachings of Jesus, such as the story of the separation of the sheep from the goats (Matthew 25:31—46) and the parable of the rich man and Lazarus (Luke 16:19—31) clearly suggest that we will be judged by our thoughts and actions—by how we treat our neighbor or whether we love God more than money, for example. One cannot read the Gospels without getting the clear message that in the day of judgment those who have been unselfish,

---

3. Barlow, *The Autobiography of Charles Darwin*, 87.

humble, and loving will fare better than those who have been greedy, arrogant, and hateful. Paul says in II Corinthians 5:10 that "we must all appear before the judgment seat of Christ, that each one may receive what is due him for the things done while in the body, whether good or bad." "Those who have done good will rise to live, and those who have done evil will rise to be condemned," Jesus says in John 5:29.

This sounds fair enough—in fact the problem is that it is too fair! Martin Luther and other protestant reformers, despairing of the hopelessness in trying to be righteous enough to be sure one has earned salvation through good works, re-discovered and emphasized the many passages in the New Testament which talk of salvation by faith. For example, Mark's version of the Great Commission (Mark 16:15-16) says, "Go into all the world and preach the good news to all creation. He who believes and is baptized will be saved, but whoever does not believe will be condemned." Or John 3:18: "Whoever believes in him is not condemned." Paul's letters especially emphasize that "by grace you have been saved, through faith . . . not by works, so that no one can boast" (Ephesians 2:8-9). This is a more comforting thought. All we have to do is believe; we need not constantly worry about whether we are good enough for God. We are not, but God forgives us because of our faith.

But what about the millions of people who have never heard of Jesus or whose only contact with him has been so superficial or even negative that we could hardly expect them to believe? Salvation by faith may be a comforting thought to me because I believe, but what if I had been a tenth century American Indian or an Aztec whose only exposure to Christianity was what Hernan Cortes brought from Spain? I could hardly be blamed or condemned for my lack of faith in Christ. Consider a person like Anwar Sadat, who braved the anger of radical Moslems in Egypt to fly to Jerusalem to make peace with Israel, knowing that it might cost him his life (it did), and who wrote in his last memoirs that "man knows by intuition that divine love was the secret behind the creation of man" and that man can get to know God "by contemplating the beauty of the flower, the greenery of the trees . . . " Why should such a man be "condemned" for lack of faith in Jesus? "Judgment" implies that some form of "justice" is done!

I have heard it argued that everyone has sinned and deserves condemnation, and that only through faith can we be forgiven and become "perfect" and therefore free from the wrath of God. It may seem unfair to us, we are told, to condemn "good" but imperfect Hindus, but God's idea of justice demands eternal punishment for even the slightest sin unless that sin is washed away by faith. Is there some theological principle involved here that only the theologians can appreciate? If they are right, then God has given us a sense of justice very different from his own and expects us to be much more merciful than he is! And if faith is the only standard by which we are to be judged, why did Jesus answer "love" every time he was asked about the most important commandment? Why did even Paul, who stressed the importance of faith more than anyone else, say (I Corinthians 13:2) "if I have a faith that can move mountains, but have not love, I am nothing." Paul also says (Romans 2:14-15) that the "Gentiles who do not have the law" will be judged by their consciences, since "the requirements of the law are written on their hearts." Peter (Acts 10:34-35) says, "I now realize how true it is that God does not show favoritism but accepts men from every nation who fear him and do what is right." If we look at the picture of God painted by the New Testament as a whole, rather than focusing on three or four isolated verses, we cannot possibly reconcile this picture with the idea of a God who would condemn entire races of people for not being Christian. It is always dangerous to base one's theology on isolated passages of the Bible; in any case, let us now look more carefully at these verses.

First, let us look again at Mark's version of the Great Commission: ". . . preach the good news . . . whoever does not believe will be condemned." Surely Jesus is talking about those who hear the good news that Jesus showed in his life and death that God loves us, and that he taught by word and example that we should love our neighbor and our enemy, and that we should be unselfish and humble in our dealings with others, and who do not like the message. He is clearly not referring to those who neither accept nor reject the gospel because it never arrives at their door.

Perhaps we can better understand what Jesus said to his disciples when he sent them out to tell the good news to the world by looking at what he said when he sent them out to teach only the Jewish people at an earlier date (Matthew 10:12-14): "As you enter the home, give it your greeting. If the home is deserving, let your peace rest on it . . . . If anyone will not welcome you or listen to your words, shake the dust off your feet when you leave that home . . . ." No one could imagine that this implied a condemnation of those to whose homes the disciples never arrived; it is simply a condemnation of those who chose to reject the good news of the love of God. Therefore, it seems reasonable to conclude that Jesus was making a similar statement in his Great Commission. In fact, Jesus may not have said "whoever does not believe will be condemned" at all, since the other gospel account (Matthew 28:16-20) of the Great Commission does not include this, and the most reliable early manuscripts do not contain Mark 16:9-20.[4]

Furthermore, let us examine Jesus' claim that "no one comes to the Father, except through me" (John 14:6) in its context. He said, "No one comes to the Father, except through me. If you really knew me, you would know my Father as well. From now on, you do know him and have seen him." In its context, it seems clear to me that Jesus is not talking about getting to heaven, but getting to know God. We can learn something about God from Nature: that he is very intelligent, for example, but not much more. Only through the life and teachings of the historical Jesus can we really begin to understand the nature of our invisible God. "Anyone who has seen me has seen the Father," he says again in verse 9.

Let us look again also at John 3:18: "Whoever does not believe stands condemned." But John continues "and this is the condemnation, that light has come into the world, but men preferred the darkness, because their deeds were evil."

Immediately after Jesus rose from the tomb, his enemies (mostly religious leaders who feared losing their power over the people) paid false witnesses to spread the rumor that his disciples had stolen his body. It was exactly as Jesus had predicted in the parable of the rich man and Lazarus: "If they do not listen to Moses and the prophets, they will not be convinced even if someone rises from the dead." I imagine these were the kinds of people who "prefer the darkness" that John was thinking about when he wrote these words. Jesus told the Pharisees, "If you were blind, you would not be guilty of sin; but now that you claim you can see, your guilt remains." (John 9:41)

Indeed, many people in the world today prefer the darkness. It is a waste of time to present evidence to them because they don't believe simply because they don't want to believe. Those who become rich selling drugs to heroin addicts, those who slaughter the innocent in their quest for power, those who sell weapons to both sides of every international dispute, corrupt political leaders— they prefer the darkness. These are the people John was talking about. But many others don't believe in Jesus because they know little or nothing about what he did and taught, and many others don't believe because the brand of Christianity to which they have been exposed bears little resemblance to the original. John was clearly not criticizing the tenth century Indian or Anwar Sadat; if the light of Christ has not come into their world, they cannot be condemned for rejecting that light.

The New Testament writers, especially Paul, often talk about salvation by faith, but—with the few exceptions examined above— always in a positive sense. The statements in the New Testament on justification by faith—"For we maintain that a man is justified by faith apart from observing the law" (Romans 3:28) and "Therefore, since we have been justified through faith, we have peace with God," (Romans 5:1) for example—when read in context are clearly not intended to condemn those who, for reasons unrelated to their preference for darkness or light, have not come to faith in Jesus, but rather to assure Martin Luther and the rest of us who have that if there is just enough of a spark of goodness in us that we are attracted to the light of Jesus' teachings, we will be forgiven, no matter how often we fall short of the righteousness we *want* to obtain. They are also intended to ensure that, as Paul puts it, no man can "boast" that he has earned his reward.

But if I cannot be blamed for rejecting the light which no missionary has brought to my village, why then should the missionary bother to come? To bring me light! Jesus said of himself, "I am the light of the world." Matthew said that Jesus fulfilled the prophecy of Isaiah: "The people living in darkness have seen a great light; on those living in the land of the shadow of death, a light has dawned." (Matthew 4:16) And how much the world needs that light today! Although I am well aware that at certain times and certain places, the church has been more of a champion of darkness than of light, I have no doubt that overall it has been a tremendous force for good in the world.

In the Great Commission, and in other places in the New Testament, we are told to share joyfully the "good news" with others. Is this the good news, that after all the trials they go through in this life, most of the world is headed—without knowing it—for an even worse place, unless they accept a Savior they have heard little or nothing about? No, I believe the good news shared by the early apostles is not that Christ offers salvation from theological problems caused by sin that they were not aware of, but salvation from a separation from their "Father in heaven" that they are very well aware of, and very real solutions to real problems caused by the sins of hatred, dishonesty, greed, envy, cruelty and corruption.

Though the very word "gospel" means "good news," the gospel some churches have been spreading is certainly not good news, and Christianity will never set the world on fire again until we start preaching good news again. That we haven't done a very good job of presenting the gospel as good news is evidenced by an advertisement that was run recently on some London buses: "There's probably no God. Now stop worrying and enjoy your life." Maybe we could express the good news of the gospel by modifying this a little: "God loves you. Now stop worrying and enjoy your life, and help others to enjoy theirs."

If we read the Bible like a law book, we may find it very confusing. For example, sometimes God's standards seem impossibly high ("Be perfect, therefore, as your heavenly Father is perfect"), while sometimes God is portrayed as much less demanding ("Come to me, all you who are weary and burdened, and I will give you rest . . . for my yoke is easy and my burden is light"). When a rich young ruler, who had kept all the commandments since his youth, asked "What must I do to inherit eternal life?" Jesus answered, "Go, sell everything you have and give to the poor, and you will have treasure in heaven." Yet he said "Today you will be with me in Paradise" to a thief who first repented of his sins and believed in Jesus as he was dying on the cross next to Jesus. To make sense of the apparently conflicting statements in the New Testament as to what God expects of us, we need to think of him, not as a disinterested judge, but as a loving father. God has high expectations for his children (he wants us to at least stop torment-ing each other), and sometimes he scolds us or even threatens us for not living up to them, but when we fall short of his standards and feel bad about it, like a father he comforts us and tells us he still loves us.

Jesus clearly taught that there will be a judgment, where those whose greed, corruption, and cruelty made their brothers' and sis-ters' lives on Earth unbearable will be punished in some manner. I wonder if the threat of punishment in the Bible might have been overstated to influence our behavior—fathers have been known to do that—and sometimes I am even tempted to think that maybe it is *entirely* there to influence our behavior, that maybe there will actually be amnesty for everyone at the time of judgment. But if there is to be amnesty (or unexpected leniency) for all, it is surely best for our stay here that we not know this yet. You may say that God should not have to resort to threats to get us to behave, but that is naive. History has shown that when people believe there will not be any punishment for evil, or any reward for good, they usually behave accordingly.

On the other hand, if there is an afterlife, would it really be fair for Hitler, Stalin, and bin Laden to be treated the same as their victims?

When we think about reward and punishment after death, about all we can do is speculate. But in the story of the prodigal son, in the Lord's prayer, and in other teachings, Jesus pictured God as "our Father in heaven," so one thing we can be sure of is that God will not punish any of his children for not acting on a command they did not understand or did not even hear. Even Earthly fathers don't do that.

5.3 THE CROSS

A central tenet of Christian theology is the idea that Jesus died "to save us from our sins." Why was it necessary for Jesus to die on the cross before God could forgive us our sins? Is God, as some claim, really so "just" that he has to punish someone for our misdeeds—if not us, then an innocent person? Is this another theological concept that only God and a few theologians can understand?

Perhaps to understand why Jesus had to die "for our sins," we need to go back into the Old Testament and look at an even stranger biblical idea: the use of animal sacrifices to "atone" for sins. Is God really so blood-thirsty that he needs to see some innocent lamb—or an innocent human—sacrificed before he can pardon us? I find it more plausible that these sacrifices were not intended to make an impression on God, but on us. As parents, when we deal with unacceptable behavior from our children, we want to impress upon them two things at the same time: that we take their misbehavior seriously and yet that we still love them. If my child is caught stealing something from a neighbor, I don't want to just say, "That's okay, I love you and forgive you—no problem." I want to make sure he understands that his behavior is very serious and must be corrected; at the same time, I also want to make sure he understands I still love him and can forgive him. Perhaps the animal sacrifices were designed to convey exactly this same dual message to God's children. Watching an animal sacrificed on an altar and being told that this was necessary for atonement for sins would surely convey the idea that, while God does forgive, our sins have serious consequences as they have resulted in the death of a poor, innocent lamb. Perhaps the sacrifice of the "lamb of God" was intended to convey even more powerfully, and finally, the seriousness of mankind's evil nature—it cost the suffering and death of an innocent man—and at the same time, the love of God—this price was paid by God himself.

We can see all the sins of mankind on display in the crucifixion story. There is the greed and corruption of the religious rulers, who were angry at Jesus for driving the merchants and money changers out of the temple, which they had corrupted from a "house of prayer" into a "den of thieves." There is dishonesty and injustice in the trumped-up charges brought against Jesus, which change from "blasphemy" to "insurrection" when the venue is changed from a Jewish court to a Roman court, where blasphemy is not an issue. There is the cowardice of Pilate, who admitted, "I find no fault in this man," while sentencing Jesus to die, because he feared antagonizing the Jews. There is the disloyalty of even his closest friends, as one betrayed him for a price and others fled or denied they knew him after he was arrested. There is hatred in the eyes of the crowd as they cry "Crucify him, Crucify him," and finally there is the extreme cruelty of human nature graphically exhibited as his persecutors put a purple robe and a crown of thorns on him and ridicule the "king of the Jews," spit at him and beat him within an inch of his life, force him to carry a heavy cross to a site outside the city, and nail his hands and feet to this cross, upon which he hangs for hours until he dies. The crucifixion story impresses us with the love of God, who came to Earth to share in our sufferings and to announce forgiveness for our sins. But no one can hear this story without also being impressed by the seriousness of those sins, by the degenerate state of a world in which even God Incarnate is treated with cruelty.

The Christian message can be condensed to two great themes: law and gospel (good news). Throughout the New Testament we find both law—that God holds out a high standard of behavior before us and takes our sins very seriously because they cause others so much pain—and good news—that God understands that we are only human and is ready and willing to forgive us every time we fall short of this standard. It has been said that the law is designed to afflict the comfortable, and the gospel to comfort the afflicted. We see these two great themes converge in the cross, which is designed to impress upon us both the seriousness of our sins and, at the same time, the love and forgiveness of "our Father in heaven." And this is exactly the impression it has left on millions of Christians throughout the years.

6

Is God Really Good?

6.1 IS GOD REALLY GOOD?

Why do bad things happen to good people? This is the question which Rabbi Harold Kushner, in his highly-acclaimed book *When Bad Things Happen to Good People*,[1] called "the only question which really matters" to his congregation. It is a question which has been asked by philosophers and ordinary human beings throughout the ages; if it is not the most-asked question, it is certainly the most passionately-asked. It was certainly the first question that occurred to me in 1987 when I was told that my beloved wife Melissa, 34 years old and the mother of our two small children (Chris and Kevin), had cancer of the nose and sinuses, and in 1990 when we discovered that the cancer had recurred. The suffering she bravely endured during those years, from the aggressive chemotherapy treatments, each of which required hospitalization for severe nausea and other side effects; from the radiation therapy; and from three major surgeries, was beyond description. Before the last surgery, during which they would remove her left eye and half of her teeth, she said, well, many people would be happy to have one eye. The cancer recurred two months after this surgery, and I was terribly depressed for many years after her death. Since

1. Kushner, *When Bad Things Happen to Good People*.

I am a pretty logical person, it never occurred to me to ask "Does God really exist?" but I certainly wondered, "Is God really good?"

Figure 6-1. Melissa Wehmann Sewell (1953-1991), with Chris

I think most people who claim not to believe in God say this not because of any shortage of evidence for design in Nature but because it is sometimes so hard to see evidence that God cares about us, and they prefer not to believe in God at all rather than to believe in a God who doesn't care.

Of course, Christians point to the life and death of Jesus as the ultimate proof that God does care about us because he came to live and suffer with us. Jesus asked the same question we have all asked at some time in our lives: "My God, my God, why have you forsaken me?" But while it is comforting to think that, despite all evidence to the contrary, God really does care about us, that still does not explain why the world God made is sometimes so cruel.

A wonderful little article in *UpReach* by Batsell Barrett Baxter entitled "Is God Really Good?"[2] contains some insights which I have found very useful into the "problem of pain,"[3] as C. S. Lewis

2. Baxter, *Is God Really Good?*
3. Lewis, *The Problem of Pain.*

calls it. I will follow Baxter's outline in presenting my own thoughts on this question, and I would like to begin with his conclusion: "As I have faced the tragedy of evil in our world and have tried to analyze its origin, I have come to the conclusion that it was an inevitable accompaniment of our greatest blessings and benefits." In his outline, Baxter lists some examples of blessings which have, as inevitable consequences, unhappy side effects. None of these points is likely to make suffering in its severest forms any easier to accept, and we may be left wondering whether these blessings are really worth the high cost. But I believe they do at least point us in the right direction.

6.2 THE REGULARITY OF NATURAL LAW

The laws of Nature which God has made work together to create a magnificent world of mountains and rivers, jungles and waterfalls, oceans and forests, animals and plants. The basic laws of physics are cleverly designed to create conditions on Earth suitable for human life and human development. Gravity prevents us and our belongings from floating off into space; water makes our crops grow; the fact that certain materials are combustible makes it possible to cook our food and stay warm in winter. Yet gravity, water, and fire are responsible for many tragedies, such as airplane crashes, drownings, and chemical plant explosions. Tragedies such as floods and automobile accidents are the results of laws of physics which, viewed as a whole, are magnificently designed and normally work for our benefit. Nearly everything in Nature which is harmful to man has also a benevolent side, or is the result of a good thing gone bad. Even pain and fear themselves sometimes have useful purposes; pain may warn us that something in our body needs attention, and without fear, we would all die young doing foolish and dangerous things or kill ourselves the first time life disappoints us.

Figure 6-2. ". . . a magnificent world, of mountains and rivers, jungles and
waterfalls . . ." (Rio Carrao, Venezuela)

But why won't God protect us from the bad side effects of
Nature? Why doesn't he overrule the laws of Nature when they
work against us? Why is he so "silent" during our most difficult
and heart-breaking moments? First of all, if we assume he has
complete control over Nature, we are assuming much more than
we have a right to assume. It does not necessarily follow that because something is designed, it can never break down. We design
cars, and yet they don't always function as designed. When our car
breaks down, we don't conclude that the designer planned for it
to break down, nor do we conclude that it had no designer; when
the human body breaks down, we should not jump to the conclusion that God planned the illness, nor should we conclude that the
body had no designer.

That we were designed by a fantastically intelligent super-
intellect is a conclusion which is easily drawn from the evidence
all around us. To jump from this to the conclusion that this creator
can control *everything* is quite a leap. In fact, I find it easy to draw
the opposite conclusion from the evidence: that this creator *cannot*,

or at least does not, control everything. Nearly everyone seems to assume that if you attribute anything to God, you have to attribute everything to God. And even if we assume he has complete control over Nature, it is hard to see how he could satisfy everyone. Your crops are dry so you pray for rain—but I am planning a picnic. It seems fairer to let Nature take its course and hope we learn to adapt. Controlling the motions of all the atoms in the world so that nothing terrible ever happens to us, so that we always get what we most need, is probably not as easy as it sounds!

In any case, what would life be like if the laws of Nature were not reliable? What if God could and did stand by to intervene on our behalf every time we needed him? We would then be spared all of life's disappointments and failures, and life would certainly be less dangerous, but let us think about what life would be like in a world where nothing could ever go wrong.

I enjoy climbing mountains—small ones. I recently climbed an 8,700 foot peak in the Guadalupe Mountains National Park and was hot and exhausted, but elated, when I finished the climb. Later I heard a rumor that the Park Service was considering building a cable car line to the top, and I was horrified. Why was I horrified? That would make it much easier for me to reach the peak. Because, of course, the pleasure I derived from climbing that peak did not come simply from reaching the top; it came from knowing that I had faced a challenge and overcome it. Since riding in a cable car requires no effort, it is impossible to fail to reach the top, and thus taking a cable car to the peak brings no sense of accomplishment. Even if I went up the hard way again, just knowing that I could have ridden the cable car would cheapen my accomplishment.

When we think about it, we see in other situations that achieving a goal brings satisfaction only if effort is required, and only if the danger of failure is real. And if the danger of failure is real, sometimes we will fail.

When we prepare for an athletic contest, we know what the rules are, and we plan our strategy accordingly. We work hard, physically and mentally, to get ready for the game. If we win, we are happy knowing that we played fairly, followed the rules, and achieved our goal. Of course we may lose, but what satisfaction would we derive from winning a game whose rules are constantly being modified to make sure we win? It is impossible to experience the thrill of victory without risking the agony of defeat. How many fans would attend a football game whose participants are just actors, acting out a script which calls for the home team to win? We would all rather go to a real game and risk defeat.

Life is a real game, not a rigged one. We know what the rules are, and we plan accordingly. We know that the laws of Nature and of life do not bend at our every wish, and it is precisely this knowledge which makes our achievements meaningful. If the rules of Nature were constantly modified to make sure we achieved our goals—whether they involve proving Fermat's Last Theorem, getting a book published, finding a cure for Alzheimer's disease, earning a college degree, or making a small business work—we would derive no satisfaction from reaching those goals. If the rules were even occasionally bent, we would soon realize that the game was rigged, and just knowing that the rules were flexible would cheapen all our accomplishments. Perhaps I should say, "if we were *aware* that the rules were being bent," because I do believe that God has intervened in human and natural history at times in the past, and I would *like* to believe he still intervenes in human affairs, and even answers prayers, on occasions, but the rules at least *appear* to us to be inflexible.

If great works of art, music, literature, or science could be realized without great effort, and if success in such endeavors were guaranteed, the works of Michelangelo, Mozart, Shakespeare, and Newton would not earn much admiration. If it were possible to realize great engineering projects without careful study, clever planning, and hard work or without running any risk of failure, mankind would feel no satisfaction in having built the Panama Canal or having sent a man to the moon. And if the dangers Columbus faced in sailing into uncharted waters were not real, we would not honor him as a brave explorer. Scientific and technological progress are made only through great effort and careful study, and not every scientist or inventor is fortunate enough to leave his mark, but anyone who thinks God would be doing us a favor by dropping a book from the sky with all the answers in it does not understand human nature very well. That would take all the fun out of discovery. If the laws of Nature were more easily circumvented, life would certainly be less frustrating and less dangerous, but also less challenging and less interesting.

Many of the tragedies, failures, and disappointments which afflict mankind are inevitable consequences of laws of Nature and of life which, viewed as a whole, are magnificently designed and normally work for our benefit. And it is because we know these laws are reliable and do not bend to satisfy our needs that our greatest achievements have meaning.

6.3 THE FREEDOM OF MAN'S WILL

I believe, however, that the unhappiness in this world attributable to "acts of God" (more properly called "acts of Nature") is small compared to the unhappiness which we inflict on each other. Reform the human spirit and you have solved the problems of drug addiction, drunk driving, war, broken marriages, child abuse, neglect of the elderly, crime, corruption, and racial hatred. I suspect that many (not all, of course) of the problems which we generally blame on circumstances beyond our control are really caused by, or aggravated by, man—or at least could be prevented if we spent as much time trying to solve the world's problems as we spend in hedonistic pursuits.

God has given us on this Earth the tools and resources necessary to construct, not a paradise, but something not too far from it. I am convinced that the majority of the things which make us most unhappy are the direct or indirect result of the sins and errors of people. Often, unfortunately, it is not the guilty person who suffers.

But our evil actions are also the inevitable result of one of our highest blessings—our free will. C. S. Lewis, in *Mere Christianity*, says,

> Free will, though it makes evil possible, is also the only thing that makes possible any love or goodness or joy worth having . . . . Someone once asked me, "Why did God make a creature of such rotten stuff that it went wrong?" The better stuff a creature is made of—the cleverer and stronger and freer it is—then the better it will be if it goes right, but also the worse it will be if it goes wrong.[4]

Why do a husband and wife decide to have a child? A toy doll requires much less work and does not throw a temper tantrum every time they make him take a bath or go to bed. A stuffed animal would be much less likely to mark on the walls with a crayon or gripe about a meal which took hours to prepare. But most parents feel that the bad experiences in raising a real child are a price worth paying for the rewards—the hand-made valentine he brings home from school and the "I love you" she whispers as she gives her mother and father a good night kiss. They recognize that the same free will which makes a child more difficult to take care of than a stuffed animal also makes him more interesting. This must be the way our Creator feels about us. The freedom which God has given to us results, as an inevitable consequence, in many headaches for him and for ourselves, but it is precisely this freedom which makes us more interesting than the other animals. God must feel that the headaches are a price worth paying; he has not taken back our free will, despite all the evil we have done. Why are there concentration camps in the world that God created? How could the Christian church sponsor the Crusades and the Inquisition? These terribly hard questions have a simple answer: because God gave us all a free will.

Jesus told a parable about "wheat and tares,"(Matthew 13) which seems to teach that the weeds of sin and sorrow cannot be eliminated from the Earth without destroying the soil of human

4. Lewis, *Mere Christianity*, 52—3.

68

freedom from which the wheat of joy and goodness also springs.[5] It is impossible to rid the world of the sorrow caused by arrogance, selfishness, and hatred without eliminating the free will which is also the source of all the unselfishness and love in the world.

If we base our view of mankind on what we see on the television news, we may feel that good and evil are greatly out of balance today; that there is much more pain than joy in the world, and much more evil than good. It is true that the amount of pain which exists in our world is overwhelming, but so is the amount of happiness. And if we look more closely at the lives of those around us, we will see that the soil of human freedom still produces wheat as well as weeds. The dark night of Nazi Germany gave birth to the heroism of Dietrich Bonhoeffer, Corrie ten Boom, and many others. The well-known play, "The Effect of Gamma Rays on Man-in-the-Moon Marigolds," is about two sisters raised by a bitter mother who suffocates ambition and discourages education. One sister ends up following the path to destruction taken by her mother; the other refuses to be trapped by her environment and rises above it. It may seem at times that our world is choking on the weeds of pain and evil, but if we look closely, we will see that wheat is still growing here.

Again we conclude that evil and unhappiness are the inevitable by-products of one of our most priceless blessings: our human free will.

6.4 THE INTERDEPENDENCE OF HUMAN LIVES

Since it is our human free will which makes our relationships with others meaningful, Baxter's third point is closely related to the second, but he nevertheless considers this point to be important enough to merit separate consideration.

Much of an individual's suffering is the direct or indirect result of the actions or misfortunes of others. Much of our deepest pain is the result of loneliness caused by the loss of the love or the

---

5. At least until the "harvest," when the wheat and weeds are fully developed, according to this parable: then the weeds *are* discarded.

life of a loved one or of the strain of a bad relationship. How much suffering could be avoided if only we were "islands, apart to ourselves." Then at least we would suffer only for our own actions and feel only our own misfortunes. The interdependence of human life is certainly the cause of much unhappiness.

Yet here again, this sorrow is the inevitable result of one of our greatest blessings. The pain which comes from separation is in proportion to the joy which the relationship provided. Friction between friends is a source of grief, but friendship is the source of much joy. Bad marriages and strained parent-child relationships are responsible for much of the unhappiness in the modern world, but none of the other joys of life compare to those which can be experienced in a happy home. Although real love is terribly hard to find, anyone who has experienced it—as I did for a few short years—will agree that the male-female relationship is truly a masterpiece of design when it works as it was intended to work.

As Baxter writes, "I am convinced that our greatest blessings come from the love which we give to others and the love which we receive from others. Without this interconnectedness, life would be barren and largely meaningless. The avoidance of all contact with other human beings might save us some suffering, but it would cost us the greatest joys and pleasures of life."

6.5 THE VALUE OF IMPERFECT CONDITIONS

We have thus far looked at suffering as a by-product of our blessings and not a blessing in itself. And certainly it is difficult to see anything good in suffering in its severest forms.

Nevertheless, we cannot help but notice that some suffering is necessary to enable us to experience life in its fullest and to bring us to a closer relationship with God. Often it is through suffering that we experience the love of God and discover the love of family and friends in deepest measure. The man who has never experienced any setbacks or disappointments invariably is a shallow person, while one who has suffered is usually better able to empathize with others. Some of the closest and most beautiful relationships occur between people who have suffered similar sorrows.

It has been argued that most of the great works of literature, art, and music were the products of suffering. One whose life has led him to expect continued comfort and ease is not likely to make the sacrifices necessary to produce anything of great and lasting value.

Of course, beyond a certain point, pain and suffering lose their positive value. Even so, the human spirit is amazing for its resilience, and many people have found cause to thank God even in seemingly unbearable situations. While serving time in a Nazi concentration camp for giving sanctuary to Jews, Betsie ten Boom told her sister, "[We] must tell people what we have learned here. We must tell them that there is no pit so deep that God is not deeper still. They will listen to us, Corrie, because we have been here."[6]

In a letter to our children composed after she realized she had lost her battle with cancer, Melissa wrote:

> While I no longer feel physically normal, . . . in an odd sort of way, I feel even more human. I have seen and felt more suffering by myself and others around me in the last few years than I probably ever would have. I have seen children still in strollers hooked up to IV chemotherapy and young children, my own children's ages, with monstrous tumors bulging from their necks. In the face of this unjust tragedy, they still had a sweet innocent smile on their faces. I have talked with young women and men my own age who are struggling with the reality of leaving their young children and spouses long before their responsibilities of parenthood are completed.
>
> I have also discovered a deepness in relationships with others that I probably never would have otherwise cultivated . . . . I have seen the compassion and love of others towards me. I have witnessed how good and true

---

6. ten Boom, *The Hiding Place*, 197.

and caring the human spirit can be. I have learned much about love from others during these times.

We might add that not only the person who suffers but also those who minister to his needs are provided with opportunities for growth and development.

C. S. Lewis concludes his essay on *The Problem of Pain* by saying, "Pain provides an opportunity for heroism; the opportunity is seized with surprising frequency."[7] As Baxter put it, "The problems, imperfections, and challenges which our world contains give us opportunities for growth and development which would otherwise be impossible."

6.6 CONCLUSIONS

In *Brave New World*,[8] Aldous Huxley paints a picture of a futuristic Utopian society which has succeeded, through totalitarian controls on human behavior and drugs designed to stimulate pleasant emotions and to repress undesirable ones, in banishing all traces of pain and unpleasantness. There remains one "savage" who has not adapted to the new civilization, however, and his refusal to take his pills results in the following interchange between "Savage" and his "civilized" interrogators:

"We prefer to do things comfortably," said the Controller.

"But I don't want comfort. I want God, I want poetry, I want real danger, I want freedom, I want goodness, I want sin."

"In fact," said Mustopha Mond, "you're claiming the right to be unhappy."

"All right then," said the Savage defiantly, "I'm claiming the right to be unhappy."

If God designed this world as a tourist resort where man could rest in comfort and ease, it is certainly a dismal failure. But I believe, with Savage, that man was created for greater things. That is why, I believe, this world presents us with such an inexhaustible

7. Lewis, *The Problem of Pain*, 157.

8. Huxley, *Brave New World*.

array of puzzles in mathematics, physics, astronomy, biology, and philosophy to challenge and entertain us, and provides us with so many opportunities for creativity and achievement in music, literature, art, athletics, business, technology, and other pursuits; and why there are always new worlds for us to discover, from the mountains and jungles of South America and the flora and fauna of Africa, to the era of dinosaurs and the surface of Mars, and the astonishing world of microbiology.

Why does God remain backstage, hidden from view, working behind the scenes while we act out our parts in the human drama? This question has lurked just below the surface throughout much of this book, and now perhaps we finally have an answer. If he were to walk out onto the stage and take on a more direct and visible role, I suppose he could clean up our act and rid the world of pain and evil—and doubt. But our human drama would be turned into a divine puppet show, and it would cost us some of our greatest blessings: the regularity of natural law which makes our achievements meaningful; the free will which makes us more interesting than robots; the love which we can receive from and give to others; and even the opportunity to grow and develop through suffering. I must confess that I still often wonder if the blessings are worth the terrible price, but God has chosen to create a world where both good and evil can flourish rather than one where neither can exist. He has chosen to create a world of greatness and infamy, of love and hatred, and of joy and pain, rather than one of mindless robots or unfeeling puppets.

Batsell Barrett Baxter, who was dying of cancer as he wrote these words, concludes, "When one sees all of life and understands the reasons behind life's suffering, I believe he will agree with the judgment which God himself declared in the Genesis story of creation: 'And God saw everything that he had made, and behold it was *very* good.'"

Bibliography

Barclay, William. *The Daily Study Bible Series.* Westminister, 1975.

Barlow, Nora, ed. *The Autobiography of Charles Darwin.* Collins, 1958.

Baxter, Batsell Barrett. "Is God Really Good?" *UpReach* Nov-Dec (1984) 19–28.

Behe, Michael. *Darwin's Black Box: The Biochemical Challenge to Evolution.* Free Press, 1996.

———. *The Edge of Evolution.* Free Press, 2007.

Cornfeld, Gaalyah and David Freedman. *Archaeology of the Bible: Book by Book.* Harper and Row, 1976.

Davies, Paul. *Other Worlds: Space, Superspace, and the Quantum Universe.* Penguin, 1997.

Dubos, Rene. *The Torch of Life.* Simon and Schuster, 1962.

Ford, Kenneth. *Classical and Modern Physics.* Xerox College Publishing, 1973.

Harrison, Edward. *Cosmology.* Cambridge University Press, 1981.

Hawking, Steven. *A Brief History of Time—From the Big Bang to Black Holes.* Bantam Books, 1988.

Heeren, Fred. *Show Me God.* Searchlight Publications, 1995.

Huxley, Aldous. *Brave New World.* Bantam Books, 1932.

Kushner, Harold. *When Bad Things Happen to Good People.* Schocken Books, 1981.

Le Conte, Joseph. *Evolution.* D. Appleton and Company, 1888.

Leggett, A.J. *The Problems of Physics.* Oxford University Press, 1987.

Lewis, C.S. *Mere Christianity.* MacMillan, 1943.

———. *The Problem of Pain.* MacMillan, 1962.

———. *Reflections on the Psalms.* Harcourt Brace, 1958.

Lohse, Bernhard. *Luther: An Introduction to His Life and Work.* Fortress, 1986.

Lönnig, Wolf-Ekkehard and Heinz-Albert Becker. "Carnivorous Plants." In *Nature Encyclopedia of Life Sciences.* Nature Publishing Group, Wiley Interscience, 2004.

Meyer, Stephen. *Darwin's Doubt: The Explosive Origin of Animal Life and the Case for Intelligent Design.* Harper One, 2013.

Rostand, Jean. *A Biologist's View.* Wm. Heinemann Ltd., 1956.

Sewell, Granville. "Entropy and Evolution," *BIO-Complexity* 2013, number 2 (2013) 1—5. (dx.doi.org/10.5048/BIO-C.2013.2).

———. "Entropy, Evolution and Open Systems." In *Biological Information: New Perspectives.* World Scientific, 2013. (dx.doi.org/10.1142/978981450 8728_0007).

———. *In the Beginning and Other Essays on Intelligent Design, 2nd edition.* Discovery Institute Press, 2015. (www.discoveryinstitutepress.com/sewell)

———. "A Mathematician's View of Evolution," *The Mathematical Intelligencer* 22, number 4 (2000) 5–7. (see article at www.evolutionnews.org, July 23, 2015).

———. *The Numerical Solution of Ordinary and Partial Differential Equations, third edition.* World Scientific, 2015.

Simpson, George Gaylord. "The History of Life." In Volume I of *Evolution after Darwin.* University of Chicago Press, 1960.

Strauss, Walter. *Partial Differential Equations, an Introduction, second edition.* John Wiley & Sons, 2008.

Tacitus. *Tacitus Annals.* Translated by Cynthia Damon. Penguin, 2012.

ten Boom, Corrie. *The Hiding Place.* Chosen Books, 1971.

Urone, Paul Peter. *College Physics.* Brooks/Cole, 2001.

Varghese, Roy, ed. *The Intellectuals Speak Out About God.* Regnery Gateway, 1984.